RESEARCH ON CHEMICAL KINETICS

Research on Chemical Kinetics

Edited by

G.E. Zaikov

Nova Science Publishers, Inc.

Art Director: Christopher Concannon
Graphics: Elenor Kallberg and Maria Ester Hawrys
Manuscript Coordinator: Roseann Pena
Book Production: Tammy Sauter, Benjamin Fung
and Michelle Lalo
Circulation: Irene Kwartiroff and Annette Hellinger

Library of Congress Cataloging–in–Publication Data
available upon request

ISBN 1-56072-094-8

6080 Jericho Turnpike, Suite 207
Commack, New York 11725
Tele. 516-499-3103 Fax 516-499-3146
E Mail Novasci1@aol.com

Printed in the United States of America

CONTENTS

PREFACE

"If you like to outlive yourself and to be honorable by posterity, then leave a virtual family and a good book."
Pythagoros (Ancient Greece)

We could say that Professor Nickolay M. Emanuel (1915-1984) was one who did follow Pythagoros. The daughter of the Professor, Olga, holds the post of Senior Research Scientist in the Department of Chemical and Biological Kinetics of the Institute of Chemical Physics (Moscow), his grandson Alexey is finishing undergraduate work in the Department of Chemistry of Moscow State University. Professor Emanuel himself was at the Head of Division of General and Technical Chemistry of the Presidium of the USSR Academy of Sciences for a long time as well as being one of the world leaders in the area of chemical kinetics. He wrote 20 books, 70 reviews, and more than 500 papers which dealt with many aspects of chemical kinetics.

The present collection was prepared by pupils of Professor N. Emanuel.

Chemical Kinetics is a science which studies chemical process, regularities and velocity as well as the mechanism of chemical transformations. The slogan of this is "Research ⇒ Development ⇒ Production."

In fact, many fundamental research results in the area of chemical kinetics were on the basis of new chemical technologies. Chemical Kinetics also is the basis of many new branches of the chemical industry and one of the essential factors in the struggle for improvement of the environment.

Here, we should remember the words of Professor Burton who starkly defined the present situation as follows: "Only one thing can happen: either people manage to reduce the amount of smoke in the air or it will cause a reduction of the Earth's population." Jacques Ives Cousteau also summed up this problem by comparing the Earth to be "a car driving alone in space with its exhaust pipe inside the cab." Finally we can draw the conclusion that chemical kinetics is a fighter against all pollution.

PROFESSOR GUENNADII E. ZAIKOV

ACTIVATION OF PEROXIDES BY CARBONYLS OF TRANSIENT METALS IN VULCANIZING MEDIA

E.G. Kabanova, N.V. Kochnev,
N.V. Yablokova, and Yu.A. Alexandrov
Research Institute of Chemistry
N.I. Lobachevskii State University
Nizhnii Novgorod

The initial stage of the interaction of tert-butyl hydroperoxide and trimethyl(tert-butyl-peroxo)silane with $Fe(Co)_5$ has been found to consist in the formation of the donor-acceptor (1:1) complex in which the peroxide acts as the electron acceptor and the transient metal carbonyl acts as the electron donor. The parameters of the complex-formation have been calculated from the PMR-spectroscopy data. The process of complex-formation increases the rate of peroxides homolytic transformation and decreases the temperature range of their degradation by 80–100°. The peroxide-carbonyl of transient metals binary systems have been shown to be possibly used as crosslinking agents in solidification of organosilicon oligomers.

It is known [1] that catalytic additives of transient metals carbonyls (TMC) accelerate the homolytic decomposition of organosilicon peroxides (OSP). For this reason, the TMC-OSP binary systems can be recommended as initiators of radical polymerization and vulcanization at the temperatures below the temperature of the OSP thermal degradation.

To study the vulcanizing power of OSP in the presence of TMC we investigated the interaction mechanism of trimethyl (tert-butylperoxo)silane (I), methyltri(tert-butylperoxo)silane (II) and tert-butyl hydroperoxide·(III) with $Fe(CO)_5$, $Mo(CO)_6$ and $W(CO)_6$ in vulcanizable siloxane (S_1), siloxysilazane (S_2) oligomers and, for comparison, in toluene and nonane. Kinetic parameters of peroxides decomposition in the presence of TMC are given in Table 1 which shows that TMCs

Table 1. Decomposition of Peroxides (0.05 mole/l) in the Presence of Carbonyls of Transient Metals (0.02 mole/l) in Various Solvents

Solvent	CMT	T,K	I	II	III
	$Mo(CO)_6$	353	0.96	48.0	-
Toluene		393	-	-	6.04
	$W(CO)_6$	393	0.45	11.9	3.0
	without additive	463	2.5		
Nonane		353	2.54	-	-
		363	7.0	-	-
	Fe(CO)	373	13.5	-	-
		393	26.8	-	6.5
Toluene-S_1	$Mo(CO)_6$	353	1.3	-	-
(I:I)	$W(CO)_6$	393	0.1	-	-
Toluene-S_2	$Mo(CO)_6$	353	-	50.0	-
(I:I)	$W(CO)_6$	353	-	10.9	-

Table 2. The Effect of $Me_3SiOOCMe_3$ and $Fe(CO)_5$ Concentration on the Initial Rate of Peroxide Decomposition in Nonane at 353 K

$C_{0,1}$mole/l, at C_0 $Fe(CO)_5$ = 0.05 mole/l	$V_0 \cdot 10^6$ mole/l · s	C_0 $Fe(CO)_5$ mole/l, at $C_{0,1}$ = 0.05 mole/l	$V_0 \cdot 10^6$ mole/l · s
0.02	2.24	0.02	2.54
0.05	6.77	0.03	3.38
0.1	13.84	0.05	6.77
		0.1	6.5
n = 1.10 ± 0.05		n = 1.1 ± 0.2	

decrease the temperature range of peroxides degradation by 80–100°.According to their ability to accelerate the decomposition of peroxides, TMCs are arranged as follows: $Fe(CO)_5$ > $Mo(CO)_6$ > $W(CO)_6$. It was found that in S_1 oligomer as well as in toluene and nonane the decomposition of peroxides in the presence of TMCs proceeds readily up to deep transformation degree (95%) in the temperature range 353–393 K. In S_2 oligomer, the decomposition of peroxide II was investigated to 75% of transformation due to difficulties arising upon isolation of peroside from crosslinked compound. Increasing rate of peroxide degradation in S_2 oligomer (Table 1) is likely to result from the presence of active silazane groupings in the solvent, these grouping being capable of coordinating the peroxide [2].

Changes in concentrations of peroxide or $Fe(CO)_5$ in the reaction mixture at the constant concentration of the second component was found to be described by the first-order equation both with respect to peroxide and TMC (Table 2). The dependence of the initial decomposition rate (v_0) of peroxide I on the $Fe(CO)_5$ concentration is of marginal character (Figure 1), which may be explained by the formation of complex between the components (1:1).

The formation of peroxides I and III complexes with $Fe(CO)_5$ was studied by the method of PMR-spectroscopy.The investigation was performed in CCl_4 solution in the temperature range 253–293 K in the conditions which prevent the reaction between the initial substances. Displacement of chemical shifts of protons of Me_3Si- and Me_3C-groups of peroxides in the presence of $Fe(CO)_5$ (Figure 2) towards the strong field coincides in terms of direction with the displacement of chemical shifts of peroxide protons in the $Me_3SnOOCM_3$-pyridine system [3]. It proves the fact that in the $Me_3SiOOCMe_3 \cdot Fe(CO)_5$ and $Me_3COOCMe_3 \cdot Fe(CO)_5$ complexes peroxides appear to be the electron acceptors, and $Fe(CO)_5$ acts as the donor of the electron. The parameters of complexing in the $Me_3SiOOCMe_3$-$Fe(CO)_5$ system are given in Table 3. The constant of the $Me_3COOCMe_3 \cdot Fe(CO)_5$ complex formation calculated at 273 K is equal to 0.20 ± 0.03, and

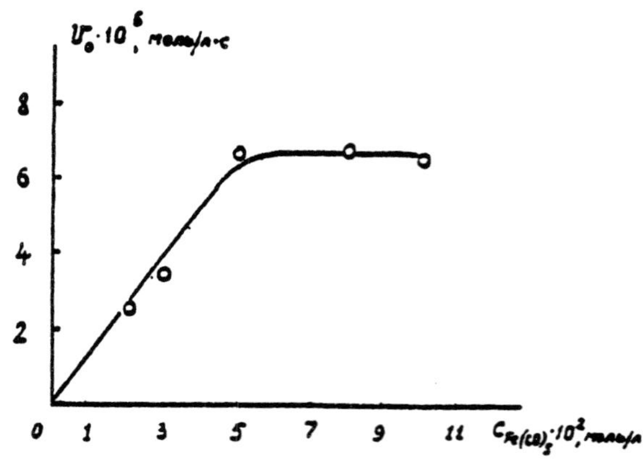

Figure 1. Dependence of the initial rate of $Me_3SiOOCMe_3$ decomposition (0.05 mole/l) on the $Fe(CO)_5$ concentration in nonane at 353 K.

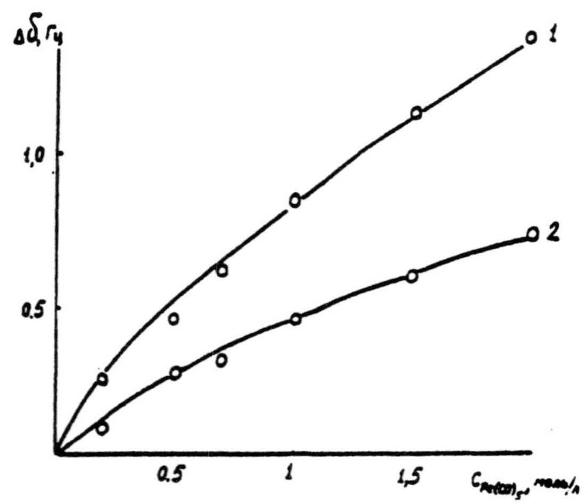

Figure 2. Dependence of the change in the chemical shift of protons of Me_3Si—(1) and Me_3C—(2) of $Me_3SiOOCMe_3$ fragments on the $Fe(CO)_5$ concentration in CCL_4 at 273 K.

- Δ H = 2.5 ± 0.3 kJ/mole. Comparison of numerical values of the $Me_3SiOOCMe_3 \cdot Fe(CO)_5$ and $Me_3COOCMe_3 \cdot Fe(CO)_5$ complexes formation ethalpies shows that organosilicon peroxide is a stronger electron acceptor than organic tert-butyl hydroperoxide with respect to $Fe(CO)_5$.

Thus it follows that the initial stage of interaction between peroxides and carbonyls of transient metals appears to be the formation of outer-sphere donor-acceptor peroxide-TMC complex (A) in which the transient metal atom enables peroxide to fill its coordination vacancies [4,5]. This penetration of the peroxide molecule into the inner coordination sphere of the transient metal atom is followed by elimination of the CO molecule [6,7].

Table 3. Complex-formation Parameters for the $Me_3SiOOCMe_3$-$Fe(CO)_5$ System in CCl_4. The Internal Reference Substance is Cyclohexane (5 vol. %)

T,K	K	-H, kJ/mole
253	0.40 ± 0.1	
263	0.25 ± 0.02	20.0 ± 2.5
273	0.23 ± 0.01	
293	0.10 ± 0.03	

A probable scheme of the initial reaction stage based on the above considerations and the geometrical parameters of the molecules of transient metals carbonyls and their complexes [8] can be represented as follows:

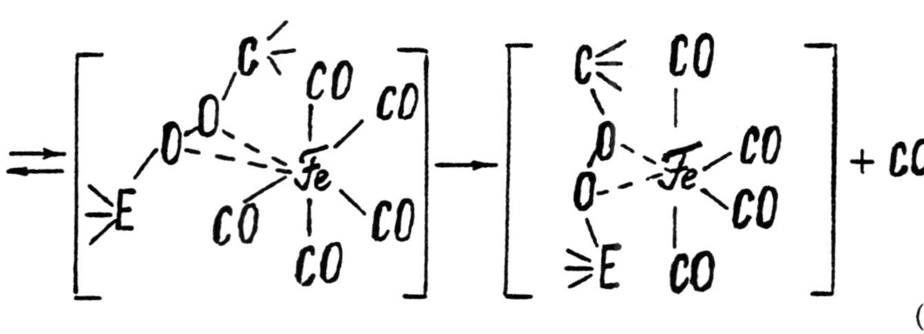

(1)

where E = C, Si

The intermediate reaction complex (B) gives rise to the conditions conducive to the transfer of d-electron from the transient metal atom onto peroxide [5,9,10] with the oxidation degree of the central metal atom increasing [10,11] and peroxide decomposing to yield radicals and ions

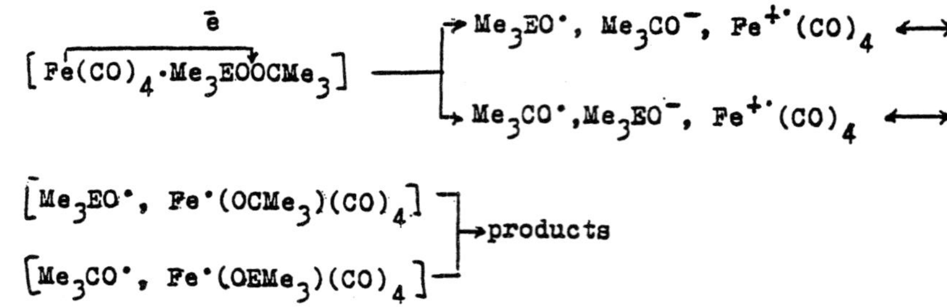

(2)

The peroxide II-TMC complex is not registered by means of PMR-spectroscopy. It has been found that mixing of peroxide II with TMC cause a reaction which results in the formation of tert-butyl hydroperoxide (TBHP) (the presence of the latter was shown by the method of GLC). A similar reaction of dimethyldi(tert-butylperoxo)silane decomposition was observed in pentaerythrite tetrabutyrate solution at 443_463 K [12]. Comparison of kinetic curves of peroxide II and carbonyl flow rate, accumulation of TBHP and tert-butanol both in proton-donor (nonane, toluene) and aproton (CCl_4) solvents shows that peroxide II activated by carbonyl gives rise to TBHC (Figure 3).

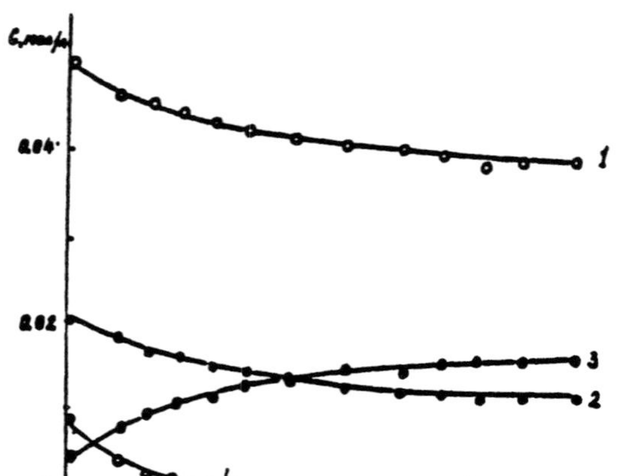

Figure 3. Kinetic curves of the $MeSi(OOCMe_3)_3$ (1) and $Fe(CO)_5$ (2) flow-rate as well as accumulation of Me_3COH (3) and Me_3COOH (4) in nonane at 295 K.

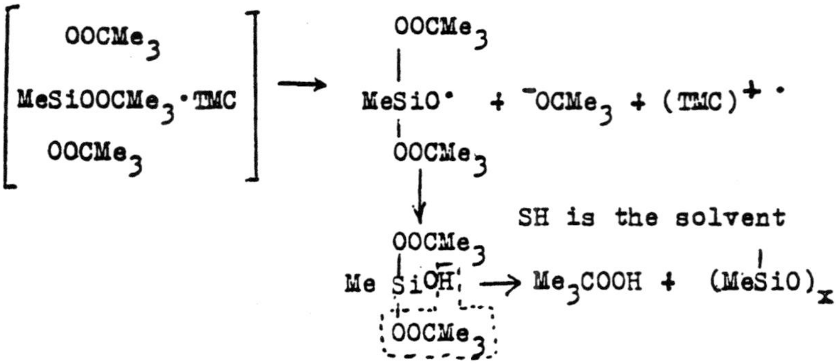

(3)

Decomposition of the peroxide II · TMC complex at low temperature prevents the determination of this complex by the method of PMR-spectroscopy

The nature of the reaction products obtained speaks for homolytic transformation of peroxides in the presence of TMC (Table 4). Alkoxyradicals $Me_3CO^{\times}$ or $Me_3SiO^{\times}$ enter the reaction of degradation and detachment of the hydrogen atom from the solvent, toluene (or nonane)

$$\begin{aligned} &Me_3\,CO^{\cdot}\,(Me_3SiO^{\cdot}) \\ &+SH \rightarrow Me_3COOH(Me_3SiOH)+S \end{aligned} \tag{4}$$

$$Me_3\,CO \rightarrow Me + Me_3\,CO \tag{5}$$

$$Me^{\cdot} + SH \rightarrow MeH + S^{\cdot} \tag{6}$$

The presence of dibenzyl as one of the reaction products indicates the formation of benzyl radicals in toluene:

$$PhCH_2 + PhCH_2 \rightarrow PhCH_2 - CH_2Ph \tag{7}$$

Dehydration of tert-butyl alcohol in the presence of transient metals carbonyls yields isobutylene:

$$Me_3\,COH \xrightarrow[H_2O]{} Me_2 = CH_2 \tag{8}$$

Table 4. Peroxides Decomposition Products (0.05 mole/l) in the Presence of Carbonyls of Transient Metals (0.02 mole/l) in Moles per Peroxide Mole

Products	$Me_3SiOOCMe$			$MeSi(OOCMe_3)_3$			$Me_3COOCMe_3$		
CH_4	0.2	0.1	0.1	0.2	0.3	0.46	0.3	0.02	0.3
C_2H_6	Traces	0.001	0.01	0.002	-	-	0.001	0.01	0.004
$Me_2C=CH_2$	0.002	0.65	0.77	0.3	0.8	0.97	0.05	2.7	2.3
Me_2CO	0.2	0.26	0.1	0.2	0.36	0.44	0.4	0.04	0.2
Me_3COH	0.6	-	-	1.3	-	0.08	1.6	0.1	-
Me_3SiOH	0.8	-	-	-	-	-	-	-	-
$(Me_3Si)_2O$	0.1	0.48	0.48	-	-	-	-	-	-
CO^*	2.0	2.2	1.2	2.4	2.0	2.0	3.0	2.5	2.2
Dibenzyl	0.3	0.28	0.14	not cald	0.6	0.36	0.6	0.04	0.1

* CO concentration is calculated in moles per CTM mole.

When peroxide I reacts with TMCs, the reaction products are found to contain $Me_3SiOSiMe_3$ formed according to the reaction of Me_3SiOH dehydration:

$$2Me_3SiOH \xrightarrow[H_2O]{} Me_3SiOSiMe_3 \qquad (9)$$

The amount of moles of carbon oxide obtained in the reaction products (Table 4) varies from 1.0 to 3 per mole of the initial TMC and negligibly increases upon transition from mono- to triperoxide. It has been found that the $Fe(CO)_5$ does not undergo decomposition in nonane and toluene in the absence of peroxides in the temperature range 353–373 K. It is probable that decomposition of transient metals carbonyls followed by displacement of CO proceeds not only according to reaction (1) but also results from degradation of complex (B) and TMC complexes with the peroxides degradation products.

It is known that peroxide vulcanization proceeds due to recombination of hydrocarbon radicals resulting from the dispatch of the hydrogen atom from the saturated hydrocarbon chain by alkoxyl and alkyl radicals resulting from peroxide degradation [13]. For this reason, the formation of dibenzyl in the reaction products (Table 4) simulates the process of crosslinking and enables quantitative assessments of the vulcanizing activity of peroxides in the presence of TMCs in the medium of siloxane and siloxysilazane oligomers.

1. EXPERIMENTAL

Synthesis of peroxides I and II was performed with the use of the techniques described in [14]. The content of the basic substance in these peroxides was equal to 99.9% (GLC, iodometry). $Fe(CO)_5$ was twice recondensed right before use under 1 mm Hg and at room temperature. $Mo(CO)_6$ and $W(CO)_6$ ("pure" grade) were used without any additional purification. Toluene and nonane were dried over metallic sodium and distilled before use. Methylvinylphenyl-siloxysilazane oligomer S_2 with MM-200-4000 contained silazane groups in addition to CH_3, CH_2= CH- and Ph-fragments.

Kinetic investigations were carried out by the ampoule method analyzing the content of peroxide oxygen in terms of time using the iodometric technique [15] (reactions of peroxides with $Mo(CO)_6$, $W(CO)_6$)

or recording the concentration of peroxide and $Fe(CO)_5$ by the method of gas-liquid chromatography (reactions of peroxides I and III with $Fe(CO)_5$).

Analysis of the reaction products was made on a Zwet-104 chromatograph with a detector according to thermoconductivity. Peroxide III, TBHC, tert-butanol, acetone, $Fe(CO)_5$ were analyzed in a glass column (200 × 0.4 cm) filled with SE-30 (5%) on N-AW-HMDS Chromaton (0.2–0.25 mm) at 333 K, dibenzyl was analyzed in a column filled with Reoplex-400 (100 × 0.4 cm) at 453 K on the same support and at the flow-rate of the helium carrier gas 50 ml/min. CO and CH_4 were analyzed in a column (200 × 0.4 cm) filled with molecular sieve (5 Å) at 353 K(T of the detector was 393 K). Analysis of isobutylene and ethane was carried out in a column (200 × 0.4 cm) filled with SEPARON CNH at 403 K. The flow-rate of the carrier gas (helium) was 60 ml/min. The PMR spectra were recorded on a "TESLA BS-487C" spectrometer with operating frequency of 80 MHz.

The experimental results were processed on an "Electronika D3-28" computer using the program executing polynomial and linear regression analyses [16] with an accuracy of 5–10%.

2. Conclusion

A probable mechanism of decomposition of organic and organosilicon peroxides in the presence of transient metals carbonyls in nonane, toluene and vulcanizing siloxane and siloxysilazane oligomers is proposed. It is found that the formation of a complex between peroxide and carbonyl of a transient metal in which peroxide acts as the acceptor of electron and carbonyl as the donor of electron increases the rate of homolytic transformation of peroxide and decreases the temperature of its degradation by 80_100°. It is shown that the binary peroxide-transient metal carbonyl system can be recommended as initiators of radical vulcanization of organosilicon oligomers.

3. References

1. **Yu.A. Alexandrov, N.P. Sluchevskaya, N.V. Yablokova, Yu.A. Kurskii, V.I. Ivanov,** and **A.S. Okunevich**, *Zh.obshch. khimii,* 1981, vol. 51, no. 2, p. 417 (Russian).

2. **E.G. Kabanova, N.V. Yablokova, Yu.A. Alexandrov,** and **O.G. Ryzhova,** *Zh.prikl.khimii,* 1990, vol. 63, no. 6, p. 1391 (Russian).
3. **V.V. Gorbatov, Yu.A. Kurskii, Yu.A. Alexandrov,** and **N.V. Yablokova,** *Zh.obshch.khimii,* 1978, vol. 48, no. 1, p. 365 (Russian).
4. **F. Basolo,** and **R. Pirson,** *Mekhanizmy Neorganicheskikh Reaktsii (Mechanism of Inorganic Reactions),* 1971, 592 pp., Mir Publishers, Moscow (Russian).
5. **B.V. Lokshin, A.G. Ginzburg,** and **E.B. Nazarova,** *Uspekhi khimii,* 1980, vol. 49, no. 2, p. 223 (Russian).
6. **I.P. Skibida,** *Uspekhi khimii,* 1975, vol. 44, no. 6, p. 1729 (Russian).
7. **E.P. Chukovskaya, R.G. Gasanov, I.I. Kandror,** and **R.Kh. Freidlina,** *Zh.Vsesoyuznogo khimicheskogo obshchestva imeni D.I. Mendeleeva,* 1979, vol. 24, no. 2, p. 161 (Russian).
8. **M. Khardittayi,** and **I. Khardittay,** *Geometriya Molekul Koordinatsionnykh Soyedinenii v Paroobrazonoi Faze (Molecule Geometry of Coordination Compounds in the Vapor Phase),* 1976, 248 pp., Mir Publishers, Moscow (Russian).
9. **D.F. Grishin,** and **V.A. Dodonov.** *Radikalnye Reaktsii Metallo-organicheskikh Soyedinenni (Radical Reactions of Organometal Compounds),* 1986, p. 60 in Mezhvuz. zbornik, Gorkii, Gorkii State University (Russian).
10. **V.S. Markevich,** and **N.Kh. Shtivel,** *Neftekhimiya,* 1973, vol. 13, no. 2, p. 240 (Russian).
11. **M.I. Farberov, G.A. Stozkhova, A.N. Bondarenko,** and **A.L. Glusker,** *Neftekhimiya,* 1970, vol. 10, no. 2, p. 219 (Russian).
12. **A.Ya. Gerchikov, S.R. Gizatullina, N.V. Yablokova, Yu.A. Alexandrov,** and **V.S. Martemyanov,** *React. Kinet. Catal. Lett.,* 1990, vol. 41, no. 1, p. 187.
13. **V.I. Myshkovski, Z.M. Nudel'man,** and **V.L. Antonovskii,** *Kauchuk i rezina,* 1988, no. 3, p. 30 (Russian).
14. **E. Buncel,** and **A. Davies.** *J. Chem. Soc.,* 1958, p. 1550.
15. **V.A. Yablokov, A.P. Tarabarina, N.V. Yablokova,** and **M.I. Stal'nova,** *Zh. obshch. khimii,* 1971, vol. 41, no. 3, p. 887 (Russian).
16. **K. Jonson.** *Chislennye Metody v Khimii (Numerical Methods in Chemistry),* 1983, p 503, Mir Publishers, Moscow (Russian).

A Novel Method for Evaluating the Energy of Electrostatic Interaction Between Surface Charged Groups of Proteins and its Application to the Kinetics of Chymotrypsin-Catalyzed Hydrolysis

L. I. Krishtalik, V. V. Topolev,
Yu. I Kharkats, and E. V. Gor
A. N. Frumkin Institute of Electrochemistry,
Academy of Sciences of the USSR,
Russia

Abstract—The shortcomings of the method for estimating the energy of interionic interaction in terms of the Tanford–Kirkwood model, with empirical correction for the accessibility of ions to the solvent, have been analyzed. It has been demonstrated that ion projection out of the protein part of the globule and decreased water dielectric constant at the protein boundary should be taken into account. An exact solution is given in terms of a model considering a layer of water with a low dielectric constant between a spherical globule and electrolyte solution. Calculations in terms of this model are based on x-ray diffraction data without any empirical corrections. The theory agrees well with the available data on the effect of chemical modification of surface ammonium ions and carboxy groups on the pK of the δ-chymotrypsin active center. The predicted increase in the pK of the α-chymotrypsin active center with increasing solution ionic strength has been proved experimentally. The effect of ionic strength and modification of surface groups on the maximum rate constant of ester hydrolysis has been discussed.

1. INTRODUCTION

Electrostatic interactions are well-known to be essential for the functioning of various biological systems. One particular aspect of this problem is the effect of charges appearing on the dissociation of ionogenic groups located close to the protein surface on the equilibrium and kinetic properties of the protein. The energy of electrostatic interaction between these groups is usually assessed in terms of the Tanford–Kirkwood (TK) model considering a protein as a sphere with a low dielectric constant ε surrounded by an electrolyte solution in a dielectric with a high ε (water) [1]. The depth of the ion immersion in the globule is critical for calculations in terms of this model.

The available data on the three-dimensional structure of proteins demonstrate, however, that most charged groups are not buried in the globule, but rather project considerably into the solution relative to the surface of the nearest part of the globule. Therefore, the commonly used version of the TK model, with charges buried in the globule, is not suitable for calculating the field of these ions.

Attempts to overcome contradictions introduced by the TK model are usually based on the idea of surface accessibility [2]. Surface accessibility (SA) is conventionally defined as the fraction of the accessible surface area of a group in protein relative to model tripeptide. It is obvious from a qualitative point of view that the more accessible a group is to the solvent, the greater the shielding effect of the solvent. Quantitatively however, it is unclear why ion field attenuation should be described by the factor (1-SA) while assuming at the same time that all ions are located exactly on the surface of the globule. Such a treatment should rather be considered as a good empirical formula as was, in fact, initially done by Gurd and coworkers [3] (for review see, for instance, [4]. Further development of the empirical parametrization of equations for interaction energy has been proposed in [5]). It is to be noted that a good empirical adaptation of a model to one particular problem, such as the analysis of protein titration curves, does not mean that the model works well with other problems, such as the analysis of electric field in the protein environment [6–8].

The attempt to introduce individual depths for each ion estimated from empirical correlation with SA rather than from x-ray diffraction evidence fails to improve the situation as well [9]. As a matter of fact, nothing but the empirical introduction of effective parameters without a distinct physical sense can be expected if one remains within the scope of the initial TK model and tries to describe the behavior of fixed ions projecting beyond the globule surface which are not considered by the model.

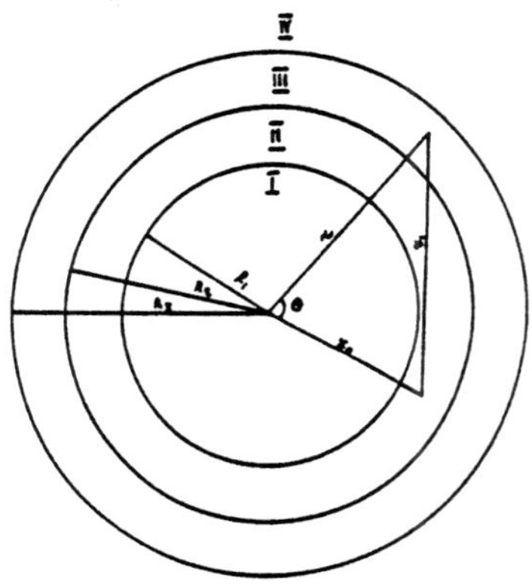

Figure 1. Model of a protein globule surrounded by a water interlayer with a small dielectric constant, aqueous electrolyte solution with a thickness equal to the Debye length and dielectric constant of the bulk water, and infinite quasimetal.

The present paper describes a physically consistent model elaborated recently by the present authors [10,11] which is free of the above drawbacks and allows the analysis of electrostatic fields solely on the basis of structural data without making resortto empirical correlations. The model is used to describe the effect of the field of surface ions on the kinetics of chymotrypsin-catalyzed hydrolysis. Discussed are the data from literature on the effect of modification of charged groups and the data of the authors on the effect of solution ionic strength.

2. MODEL

As a first step in constructing the theory one could consider the same model of two dielectric media as that utilized by the TK model, but with admitting the possibility of ions being located not only inside, but also outside the globule. This problem has been solved [10], but the agreement between the corresponding formulae and the experiment proved to be insufficient. The reason is, clearly, that the model of two dielectric media fails to describe the physical reality adequately. Indeed, it has been conclusively demonstrated by various means (see [12–14] for review) that the layer of water about one monolayer thick that borders on the protein surface differs substantially from bulk water in its properties, such as for instance dielectric relaxation time [15,16]. It is clear, therefore, that the model must take these differences into account, especially as the exposed surface ions are located just in this layer so that its dielectric properties should have a significant effect on their interaction.

Direct experimental evidence as to the magnitude of the effective dielectric constant of this layer, which is necessary for our calculations, is lacking. The measurement of the dielectric constant of weakly hydrated proteins is strongly distorted by the contribution of ionic, in particular protic, relaxation and gives therefore overestimated values. Anyway, the data available [17] indicate that ε of the surface water is by about an order of magnitude smaller than that of the bulk. On the other hand, this value can be estimated using independent data from other sources. Thus, as established from data on the capacitance of the electric double layer on hydrophobic and moderately hydrophilic electrodes, there exists an essentially ion-free 3 Å thick (about one monolayer) dielectric interlayer in the electrode-solution interface with a dielectric constant of ~ 6 (see, for instance, [18]). This value is close to 5, the infrared dielectric constant of water which, according to the modern theory of dielectric media with spatial dispersion, governs largely the short-range dielectric response [19]. It is to be noted that such a low e is not due, as often believed, to dielectric saturation in the field of the electric double layer. The actual reason is that the capacitance of the dense double layer and, consequently, the dielectric properties of the water interlayer are little affected by changes in the electrode charge and have similar magnitudes even for non-charged metal. Therefore, the point is that the boundary water monolayer has indeed particular dielectric properties [20].

In view of the above stated we considered the following model (Figure 1). A spherical globule of the radius R_1 and dielectric constant ε_1 (in the following calculations $\varepsilon_1 = 4$) is surrounded by a spherical layer with a thickness of $R_2 - R_1$ and $\varepsilon_2 = 6$. The latter is surrounded by electrolyte solution. As far as the density of charged groups on the protein surface is usually quite high and in view of inevitable approximations of the model (a sharp boundary between regions with high and low ε, independence of the layer thickness from the position of the charged group with respect to the globule boundary) we did not introduce discrete shielding of each particular charge by the ionic atmosphere, which in itself is an approximation and would result in serious mathematical complications. We used, instead, a description which proved to be a good approximate model of a diffuse electric double layer. With this end in view we introduced an interlayer of thickness $R_3 - R_2$ equal to the Debye shielding length. Solution beyond this layer was considered to have infinite dielectric constant ($\varepsilon_4 = \infty$) that is, was treated as a quasimetal. This means that the electrolyte entirely shields the electric field of charges at distances greater than the Debye length. In such a model electrolyte ions do not penetrate into interlayer II (this is unfavorable in view of a great loss of the solvation energy due to the low ε_2), and so it is not necessary to introduce the distance of the closest approach of ions.

As shown in Fig. 1, r_0 and r denote the distances from the center of the globule to the ion and to the point whose potential is to be determined, respectively. θ denotes the angle between the respective vectors. r_1 is the distance between the ion and the point in question. Both the ion and the point can be located in any region (considering their position in region IV is senseless since in this case the field is equal to zero by definition).

The solution of the electrostatic problem was obtained, as usual, given spherical symmetry in the form of Legendre series expansion [21]. Corresponding equations are given in the Appendix. Our solution reduces to the results of the TK model (or its analogs) at zero thickness of the dielectric interlayer $R_2 - R_1$.

The existence of a dielectric interlayer with $\varepsilon_2 = 6$ diminishes shielding quite significantly. Thus, the charge interaction energy estimated in terms of our model exceeds that estimated using the TK model by more than ten percent if both charges are inside the globule (though not very deep), but if one of the charges is on the surface of the globule or projects from it the difference is several-fold.

3. Application of the Model

The application of the spherical model to a real globule meets, naturally, a lot of difficulties. In the spherical model, the position of each ion is given by its distance from the center and surface of the sphere, and the mutual position of ions by the distance (or angle) between them. For a real globule, one can retain not all of these parameters, and so some of them should be given priority. The calculations were done as follows.

The center of the globule was assigned using a spatial model of the protein, and distances between the ions and the surface (positive or negative) were estimated. The position of each ion was given, therefore, by its distances from the center and surface. In the spherical model, the sum of these values is equal to the radius of the sphere for any point. In the case of a non-spherical globule such sums differ from each other that is, an individual sphere radius should, formally, be assigned to each ion. In our model each charge—test point pair was assigned a sphere of the radius equal to the arithmetic mean of the individual radii. The distance from the surface was taken as it is, and that from the center of the sphere estimated. In such a construction the centers of the sphere and globule do not coincide for each pair, and angles between the respective pairs of vectors are not equal. At the same time, the radius of the sphere is reasonable (variation within 10%), and the distances between the points and surface and between the points themselves remain unchanged, which we believe to be more important.

As far as the globule surface is not smooth, the estimation of the distance between the ion center and the surface is not unambiguous. When estimating this distance account was taken not only of the distance of the ion from the surface that forms the van der Waals envelope of the nearest to the ion surface protein atoms, but also of the overall structure of the adjacent part of the globule. Thus, the effective distance was increased in the case of a hollow and decreased in the case of a bump. Analyzed were versions with different sets of ion distances from the surface including the maximum and minimum estimates.

EFFECT OF CHEMICAL MODIFICATION OF D-CHYMOTRYPSIN SURFACE GROUPS ON THE KINETICS OF ESTER HYDROLYSIS

The modification of the surface groups of d-chymotrypsin shifts the pK of the active center by + 1 and - 1 upon the modification of amino and carboxy groups, respectively [22_24]. Although results reported by the above cited authors are similar, our calculations will primarily be referred to the data of Valenzuela and Bender [22] since in this study the groups that undergo modification have been identified more definitely.

Although Valenzuela and Bender [22] studied δ-chymotrypsin, we used x-ray coordinates reported for α-chymotrypsin [25] taking into account the absence of the terminal residues 146 and 149 since the dipeptide 147–148 is not excised in δ-chymotrypsin. Also ignored was the terminal Ile 16 since it did not undergo modification [22]. Strictly speaking, the charges do not disappear upon modification according to [22] (succinylation of amino groups and amidation of carboxy groups with ethylenediamine) since new charged groups appear. The latter however are so far away from the protein surface (they are located, in fact, in region IV) that they should not contribute to the total potential induced by surface ions. The calculations were based on the following values of parameters which we believed to be the most reasonable: $\varepsilon_1 = 4$, $\varepsilon_2 = 6$, $\varepsilon_3 = 80$, $R_2 - R_1 = 3$ Å, $R_3 - R_2 = 10$ Å (this corresponds to the Debye length for 0.1 N KCl [22]). Hydrolysis catalyzed by serine proteases involves proton transfer to the imidazole (Im) of His 57 so that the effective pK of the active center depends on the dissociation of ImH^+.

The potential induced by protonated amino groups in the center of the ImH^+ ion of His 57 was calculated as + 0.059–+ 0.061 V depending on the estimated distances of individual ions from the globule surface. The disappearance of such a potential upon amino group modification corresponds to ΔpK = 1, which agrees well with experimental data on modification.

With the modification of anions, the situation is somewhat more complicated. It has been reported that only 12 of the 14 available carboxy groups are subject to modification [22]. We believe that the two intact residues are Asp 64 and Glu 70 which are located in hollows and are therefore hardly accessible to the modifying agent. With their contribution ignored, the potential induced on His 57 by the rest of the carboxyls* would be - 0.52— 0.074 V. Its neutralization upon modification would result in a pK shift of - 1.05 ± 0.15, which again agrees with the data available from literature [22].

It would be of interest to consider the effect of modification on the catalytic constant of enzymatic hydrolysis. Experimentally it has been demonstrated that the modification of amino groups increases the maximum deacylation rate constant by a factor of 1.5 [22]. In the one-proton mechanism, the rate-limiting step involves the formation of the ImH^+ cation and a tetrahedral derivative with a negatively charged oxygen (the former carbonyl oxygen of the substrate). The change in the field resulting from the modification of charged groups should make the formation of the intermediate containing the ionic pair $ImH^+...C—O^-$ energetically less or more favorable.

We calculated that the cationic groups of the enzyme which are subject to modification induce a positive potential on the Im of His 57 which is by 0.015–0.020 V higher than that on the oxygen. This means that modification should stabilize the ionic pair corresponding to the formation of the tetrahedral derivative, the gain in energy being sufficient to account for the whole increase of the reaction rate. For example, in one of the cases considered modifiable cations induce a potential of + 0.059 V on Im and + 0.043 V on the oxygen, and the ionic pair is stabilized by 0.016 eV upon modification. Given the normal magnitude of the Broensted factor $\alpha = 1/2$, such a gain in the energy of an elementary step corresponds to an activation energy decrease by 0.008 eV that is, to a 1.4-fold increase of the rate constant.

As to the modification of surface anions, comparison with the experiment is less reliable since pH plateau was not reached [22] due to a sharp increase of the Michaelis constant. Calculations predict an energy gain of about 0.02 eV which would produce about the same acceleration as in the previous case. In the experiment the rate constant was found to increase by only several percent.

* Except for Pro 8 whose position is unknown, but obviously too remote to induce any significant potential.

5. Effect of Ionic Strength of Solution on the Kinetics of Chymotrypsin-catalyzed Hydrolysis

Changes in the ionic strength of solution affect the shielding of the field of charged groups by electrolyte ions. Although calculations did not predict the expected effect to be significant, its experimental proof still seemed possible, which prompted the study described below.

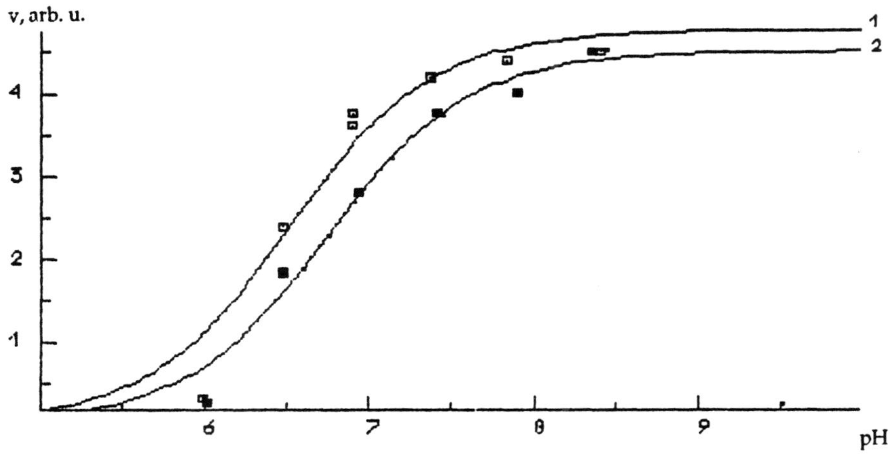

Figure 2. pH-dependence of α-chymotrypsin-catalyzed Bz-Tyr-OEt hydrolysis in the presence of 0.01 M (1) and 0.1 M (2) KCl.

6. Materials and Methods

α-Chymotrypsin (Leningrad factory of medicinal preparations, USSR) was desalted on a Sephadex G-25 column and lyophilized by a conventional technique; benzoyl tyrosine ethyl ester (Bz-Tyr-OEt) was obtained from Calbiochem, U.S.A.; NaOH was purchased from Chemapol, Czechoslovakia; acetonitrile (for photometry) and KCl (recrystallized and calcined) were products made in the U.S.S.R.

The stock solution of α-chymotrypsin (1 mg/ml) was prepared extempore. Solutions of required concentrations were prepared by appropriately diluting the stock solution. α-Chymotrypsin was dissolved in 0.001 N HCl since under these conditions the enzyme possesses the greatest stability [28]. The stock solutions of Bz-Tyr-OEt were prepared in acetonitrile. Test solutions (0.9 to 1.8 mM) contained 2 vol. % of acetonitrile.

0.01 M solution of NaOH used as a titrant was prepared by dissolving solid NaOH in water. KCl concentrations were 1.0, 0.1 and 0.01 M. The final volume of the test solution in the cell was 5 ml.

Kinetic measurements were carried out using a T-108 titrator (Gomel factory of measuring instruments, U.S.S.R.). The reaction rates were evaluated from the initial parts of the kinetic curves at a conversion of 10–15%. Prior to the addition of chymotrypsin, substrate solution was equilibrated at 21 ± 0.1° C and bubbled with nitrogen. The reaction was started by the addition of 0.1 ml of chymotrypsin solution. The reaction mixture was kept under nitrogen atmosphere.

pH was measured to an accuracy of 0.01–0.03. K_m at pH 8.5 in the presence of 0.1 M KCl was estimated as 0.031 mM, which is substantially less than the substrate concentrations employed. The same K_m at pH 8.5 was obtained in the presence of 0.01 M KCl. As far as at lower pH K_m is still smaller [29] no allowance for K_m was made.

To evaluate pK, data obtained in several series of measurements were subjected to optimization in terms of the function $F = \sum_{n=1}^{N} (y_i - Y_i)^2$,

where N is the number of points, y_i is the reaction rate and Y_i is the value determined for the same pH in terms of the formula

$$Y = v_m/(1+10^{pK-pH})$$

The data were optimized for pK and the maximum rate, v_m, with y_i corresponding to the rate at a given pK which is proportional to the concentration of the neutral form of Im in His 57.

7. Discussion

Because the effects under study are small it proves to be more reasonable to compare the data of a single series obtained on the same day and with the same portions of chymotrypsin and substrate solutions at different pH and KCl concentrations rather than deal with data averaged over several experimental series. One of such series is presented in Figure 2. The curve for 0.1 M KCl is distinctly shifted towards higher pH relative to that for 0.01 M KCl. The magnitude of the shift is 0.22 pH units as determined from the parameters of the optimized curves. In other similar experimental series where pK varied within ± 0.1, the difference between the curves of the same series amounted to 0.20–0.33, that is, was virtually the same as in Figure 2. The difference in pK between 1.0 M and 0.1 M KCl solutions was somewhat smaller and equal to 0.09 (see Figure 3).

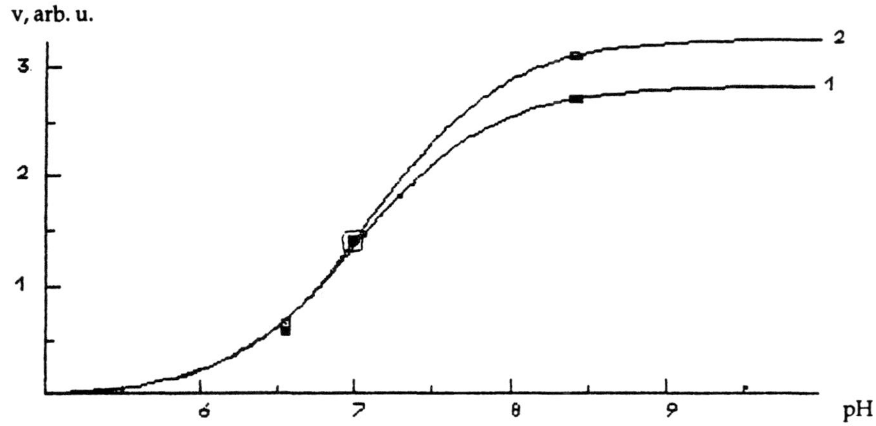

Figure 3. pH-dependence of α-chymotrypsin-catalyzed Bz-Tyr-OEt hydrolysis in the presence of 0.1 M (1) and 1.0 M (2) KCl.

The data obtained were fitted to equations for the titration of an acid with a single and constant pK. The fit proved to be reasonably good for the greatest part of the curve, but at low pH experimental values were systematically smaller than predicted ones (this is clearly seen from Figure 2 for the data obtained at pH $\simeq$ 6). Such an effect appears to arise from the distorting influence of another charged group whose dissociation is essentially completed at pH slightly above 6. However, the greatest part of the curve is reasonably close to experimental points and the magnitude of the pK shift cannot depend on the effect under consideration to any significant extent despite the fact that the shape of the fit and, consequently, pK values can be somewhat distorted.

Now compare the calculations with experimental data. The calculations were based on the same parameters (dielectric constants, shielding radius) as in the previous section. In contrast to the previous problem, which dealt with the effect of the field of only individual surface ions, in this case variation of ionic strength affects the shielding of all field sources including ionized groups, protein dipoles and the imidazolium ion of His 57 localized in the active center. Considered as ionized groups were all ammonium and guanidinium cations and carboxylate anions, both surface and internal. Tentative calculations demonstrated that the change in potential induced by an internal dipole upon ionic strength change from 1.0 M to 0.1 M did not exceed 0.0002 V and normally amounted to 10^{-2}–10^{-3} mV. In view of this fact account was taken of only those fields induced by the charged groups of the enzyme, while those of dipoles were ignored. As to the ImH^+ ion of His 57, variation of ionic strength affects its interaction with the environment and, in fact, changes its solvation energy. This effect can be evaluated from the change in the potential of ImH^+ ion image in the center of this ion $(1/2(\varphi_I$ - $\varphi_q)e$, for designations see Appendix).

Calculations show that both the change in the shielding of ionized groups and the change in ImH^+ solvation energy are of the same order. Ionic strength change from 0.01 to 0.1 M is predicted to result in a pK shift of 0.17–0.20, and further increase from 0.1 to 1.0 M in a shift of 0.06–0.11. Thus, the prediction is accurate to the experimental error taking into account not very great variation arising from the variation of the parameters.

Besides affecting the pK of the active site the shielding of electrostatic fields by electrolyte influences the maximum rate that is, k_{cat}. Ionic strength increase from 0.01 M to 0.1 M decreases k_{cat} by only 3–8%, which practically falls within the experimental error. Further increase of ionic strength from 0.1 M to 1.0 M increases k_{cat} (by 15% for the series presented in Figure 3). This is consistent with a value of 20% obtained earlier in this laboratory [31]. With the ethyl ester of acetyl tyrosine, k_{cat} increases by about 40%. Similar results have been reported by Kahana and Shalitin [32], although the authors did not study the pH dependence as a function of ionic strength.

This effect may also be due to a change of the field in the globule and of solvation energy resulting from changes in the shielding action of solution. The activation energy of charge transfer is presently believed to depend on the enthalpy of intermediate product formation, energy of reagent approach and reorganization energy of the medium [33–36]. The intermediate state at the deacylation step is commonly viewed as a tetrahedral derivative

$$\begin{array}{ll} (His\,57)-\mathrm{Im}H^{+} & \quad\;\; O-(Ser\,195) \\ & HO-\overset{|}{\underset{|}{C}}-O^{-} \\ & \quad\;\; R \end{array}$$

with the formation of an ionic pair carrying a positive charge on the Im of His 57 and a negative one on the former carbonyl oxygen of the ester. The charged groups of the protein induce fields in the globule which stabilize the ImH^{+} state of His 57 (by - 0.71 V in one of the calculation variants, with - 0.54 V contributed by the COO^{-} group of Asp 102) and make the appearance of a negative charge on the carbonyl oxygen unfavorable (-0.41 V, including - 0.14 V contributed by Asp 194). As a result, the ions field stabilizes the ionic pair (gain in energy is about 0.3 eV) and facilitates the formation of the tetrahedral derivative. This stabilization effect is largely due to the Asp 102 ion, and we believe this to be the reason for the existence of the so-called catalytic triad Asp...His...Ser in all serine proteases [37]. Stabilization by the field of Asp 102 is, however, an essential but not the only factor facilitating the formation of the ionic pair in the medium with a low dielectric permittivity. Also essential is the field of polar, primarily peptide, groups of the protein whose potentials on the carbonyl oxygen and imidazolium ion are estimated as + 0.82 V and + 0.17 V, respectively (a gain in energy of 0.65 eV) [30,37].

The shielding of the fields of charged groups and change in the intrinsic solvation energy of the ionic pair (the tetrahedral derivative) was calculated in the same way as above. In this case account was additionally taken of the effect of ionic strength on the Coulomb interaction between ImH^{+} and C—O^{-}. Although not all components changed favorably on increasing the ionic strength, the net gain in energy was 6–7 meV on going from 0.1 M to 1.0 M and ~ 5 meV on going from 0.01 M to 0.1 M solution. This corresponds (at $\alpha \approx 0.5$) to reaction rate increase by 15% and 10%, respectively. As far as the calculated effects represent small changes of large values their agreement with experiment can be considered satisfactory.

8. Appendix

The solution of the Poisson equation for the model depicted in Figure 1. has the following form

$$\varphi_I = \varphi_q^i + \sum_{n=0}^{\infty} C_n^i \,(r/R_1)^n \, P_n(\cos\theta)$$

$$\varphi_{II} = \varphi_q^i + \sum_{n=0}^{\infty} [A_n^i I^n + B_n^i / I^{n+1}] \, P_n(\cos\theta)$$

$$\varphi_{II} = \varphi_q^i + \sum_{n=0}^{\infty} [D_n^i I^n + E_n^i / I^{n+1}] \, P_n(\cos\theta)$$

where φ_I , φ_{II} and φ_{III} are potentials induced by the charge in the first, second and third phases, respectively; $P_n(\cos\theta)$ is Legendre's polynomial of degree n; $A_n^i, B_n^i, C_n^i, D_n^i$,, and E_n^i are factors depending on the position of the charge in a particular phase *i*.

If the test point and the charge are located in different phases, then $\varphi_q^i = 0$. Otherwise $\varphi_q^i = q/\varepsilon r_1$, where *q* is the magnitude of the charge and ε dielectric constant of the phase. In some instances it is reasonable to expand φ_q^i into a series. In this case

$$\varphi_q^i = (\varepsilon r) \sum_{n=0}^{\infty} (r/r_0)^n \, P_n(\cos\theta)$$

if $R < r_0$ and

$$\varphi_q^i = (q/\varepsilon r) \sum_{n=0}^{\infty} (r_0/r)^n \, P_n(\cos\theta)$$

if $r > r_0$.

If the charge is located in the first phase expressions for the factors assume the following form

$$A_n^I = q(2n+1) r_0^n / [n(\epsilon_1 - \epsilon_2) R_1^{2n+1} - (n\epsilon_1 + (n+1)\epsilon_2) R_2^{2n+1} P]$$

$$B_n^I = -A_n^I R_2^{2n+1} P$$

$$C_n^I = -q r_0^n / \epsilon_1 R_1^{n+1} - A_n^I (R_2^{2n+1} P - R_1^{2n+1}) / R_1^{n+1}$$

$$D_n^I = A_n^I R_2^{2n+1} (1 - P) / (R_2^{2n+1} - R_3^{2n+1})$$

$$E_n^I = -A_n^I R_2^{2n+1} R_3^{2n+1} (1 - P) / (R_2^{2n+1} - R_3^{2n+1})$$

The respective expressions for the second and third phases are

$$A_n^{II} = -\frac{q[n(\epsilon_1 - \epsilon_2)R_1^{2n+1} - (n\epsilon_1 + (n+1)\epsilon_2)r_0^{2n+1}]}{\epsilon_2 r_0^{n+1}[n(\epsilon_1 - \epsilon_2)R_1^{2n+1} - (n\epsilon_1 + (n+1)\epsilon_2)R_2^{2n+1}P]}$$

$$B_n^{II} = -qr_0^n/\epsilon_2 - A_n^{II}R_2^{2n+1}P$$

$$C_n^{II} = q(R_1^{2n+1} - r_0^{2n+1})/\epsilon_2 r_0^{n+1}R_1^{n+1} + A_n^{II}(R_1^{2n+1} - R_2^{2n+1}p)/R_1^{n+1}$$

$$D_n^{II} = A_n^{II}R_2^{2n+1}(1 - P)/(R_2^{2n+1} - R_3^{2n+1})$$

$$E_n^{II} = -A_n^{II}R_2^{2n+1}R_3^{2n+1}(1 - P)/(R_2^{2n+1} - R_3^{2n+1})$$

$$A_n^{III} = (q/r_o^{n+1}2n + 1)(R_3^{2n+1} - r_0^{2n+1})(n\epsilon_1 + (n+1)\epsilon_2)R_2^{2n+1}/\{\epsilon_3[nR_2^{2n+1}$$
$$+(n+1)R_3^{2n+1}][(n\epsilon_1 + (n+1)\epsilon_2)R_2^{2n+1} + (n+1)(\epsilon_1 - \epsilon_2)R_1^{2n+1}]$$
$$+ \epsilon_2 n(R_3^{2n+1} - R_2^{2n+1})[(n\epsilon_1 + (n+1)\epsilon_2)R_2^{2n+1} + (n+1)(\epsilon_1 - \epsilon_2)R_1^{2n+1}]\}$$

$$B_n^{III} = A_n^{III}n(\epsilon_2 - \epsilon_1)R_1^{2n+1}/(n\epsilon_1 + (n+1)\epsilon_2) \qquad C_n^{III} = A_n^{III}R_1^n\epsilon_2(2n+1)/(n\epsilon_1 + (n+1)\epsilon_2)$$

$$D_n^{III} = \frac{qR_2^{2n+1}}{r_0^{n+1}\epsilon_3(R_3^{2n+1} - R_2^{2n+1})} - A_n^{III}\frac{(n\epsilon_1 + (n+1)\epsilon_2)R_2^{2n+1} + n(\epsilon_2 - \epsilon_1)R_1^{2n+1}}{(n\epsilon_1 + (n+1)\epsilon_2)(R_3^{2n+1} - R_2^{2n+1})}$$

$$E_n^{III} = -qr_0^n/\epsilon_3 - D_n^{III}R_3^{2n+1}$$

In these equations the factor P is given by

$$P = \frac{n(\epsilon_3 - \epsilon_2)R_2^{2n+1} + (n\epsilon_2 + (n+1)\epsilon_3)R_3^{2n+1}}{((n+1)\epsilon_2 + n\epsilon_3)R_2^{2n+1} + (n+1)(\epsilon_3 - \epsilon_2)R_3^{2n+1}}$$

When performing computer calculations it is convenient to use the recurrence formula for Legendre's polynomial

$$(n+1)\,P_{n+1}(x) = (2n+1)\,xP_n(x) - nP_{n-1}(x)$$

starting with the expressions $P_0(x) = 1$ and $P_1(x) = x$.

The convergence of the above series depends significantly on the position of the test point. In particular cases considered in the present study the number of terms varied from 50 to 500.

9. References

1. **C. Tanford,** and **J. G. Kirkwood**. *J. Amer. Chem. Soc.* Vol. 79, No. 20, 1957, 5333 p.
2. **B. Lee,** and **F. M. Richards**. *J. Molec. Biol.* Vol. 55, No. 3, 1971, 379 p.
3. **S. Shire, G.I.H. Hanania,** and **F.R.N. Gurd**. *Biochemistry,* Vol. 13, No. 1, 1974, 2 p.
4. **J. B. Matthew**. *Annu. Rev. Biophys. Biophys. Chem.* Vol. 14, 1985, 387 p.

5. **V. Z. Spassov, A. D. Korshikov,** and **B. P. Atanasov**. *Bioch. Bioph. Acta, Protein Str.* Vol. 99, No. 1, 1989, 1 p.
6. **J. B. Matthew, P. C. Weber, F. R. Salemme,** and **F. M. Richards**. *Nature,* Vol. 301, No. 5896, 1989, 169 p.
7. **I. T. Weber,** and **T. A. Steitz**. *Proc. Nat. Acad. Sci. U.S.A.* Vol. 81, No. 13, 1984, 3973 p.
8. **J. B. Matthew,** and **F. M. Richards**. *Biochemistry* Vol. 21, No. 20, 1982, 4989 p.
9. **T. Imoto**. *Biophysic. Journ.* Vol. 44, No. 3, 1983, 293 p.
10. **L. I. Kirshtalik, V. V. Topolev,** and **Yu. I. Kharkats**. *Biofizika,* Vol. 36, No. 2, 255 p. 1991 Russian.
11. **L. I. Krishtalik,** and **V. V. Topolev**. *Biofizika,* Vol. 35, No. 5, 1990, 746 p., Russian.
12. *Water in Polymers* (S. P. Rowland, ed.) Washington, D.C., American Chemical Society 1980.
13. **J. T. Edsall,** and **H. A. McKenzie**. *Adv. Biophys.* Vol. 16, 1983, 53 p.
14. **W. Saenger**. *Annu. Rev. Biophys. Biophys. Chem.* Vol. 16, 1987, 93 p.
15. **R. Pethig,** and **D. B. Kell**. *Phys. Med. Biol.* Vol. 32, No. 8, 1987, 933 p.
16. **Yu. I. Khurgin, V. D. Nissenbaum, E. V. Belyakov, T. L. Perepechkina,** and **V. I. Yakerson**. *Dokalady Akad. Nauk SSSR* Vol. 299, No. 4, 1988, 929 p.
17. **J. Eden, P. R. C. Gascoyne,** and **R. Pethig**. *J. Ch. Soc. Faraday Trans.* Vol. 76, No. 2, 1980, 426 p
18. **E. Gileadi, E. Kirowa-Eisner,** and **J. Perciner**. *Interfacial Electrochemistry. An Experimental Approach. Reading etc.,* Addison-Wesley Publ. Corp. 1975, 525 p.
19. **A. A. Kornyshev**. *Electrochimica Acta* Vol. 26, No. 1, 1981, 1 p.
20. **L. I. Krishtalik**. *Elektrokhimiya* Vol. 13, No. 12, 1977, 1890 p., Russian.
21. **J. A. Stratton**. *Electromagnetic Theory.* New York and London. McGraw-Hill, 1941, 615 p.
22. **P. Valenzuela,** and **M. L. Bender**. *Biochim. Biophys. Acta* Vol. 250, No. 3, 1971, 538 p.
23. **V. P. Torchilin, I. L. Reizer,** and **E. I. Chazov**. *Bioorganicheskaya Khimiya* Vol. 2, No. 9, 1976, 1252 p., Russian.
24. **B. H. J. Hofstee**. *Biochem. Biophys. Res. Comm.,* Vol. 41, No. 5, 1970, 1141 p.
25. **J. J. Birktoft,** and **D. M. Blow**. *J. Mol. Biol.,* Vol. 68, No. 2, 1972, 187 p.
26. **A. Fersht**. *Enzyme Structure and Mechanism,* Reading and San Francisco, W. H. Freeman and Co., 1977, 371 p.

27. **M. Dixon,** and **E. C. Webb**. *Enzymes*. New York, San Francisco, Academic Press, Logman, 1979, 1116 p.
28. **P. E. Wilcox** in *Methods in Enzymology,* Vol. 19 (S. P. Colowick and N. O. Kaplan, eds.), New York, London: Academic Press, 1970, 79 p.
29. **K. V. Antonov**. *Khimiya proteoliza* (Proteolysis Chemistry). Moscow, Nauka, 1983, 368 p., Russian.
30. **L. I. Krishtalik,** and **V. V. Topolev**. *Molekularnaya Biologiya,* Vol. 18, No. 4, 1984, 892 p. Russian.
31. **D. E. Khoshtariya,** and **V. V. Topolev**. *Bioorganicheskaya Khimiya* Vol. 6, No. 3, 1980, 455 p., Russian.
32. **L. Kahana,** and **Y. Shalitin**. *Israel Journ. Chem.,* Vol. 12, No. 2, 1974, 573 p.
33. **R. R. Dogonadze,** and **A. M. Kuznetsov**. *Progr. Surf. Sci.,* Vol. 6, 1975, 545 p.
34. **R. R. Dogonadze,** and **A. M. Kuznetsov** in *Comprehensive Treatise of Electrochemistry* (J. O. M. Bockris, B. E. Conway, E. Yeager, S. U. M. Khan and R. E. White, eds.) Vol. 7, New York, London Plenum Press, 1983, 1 p.
35. **L. I. Krishtalik**. *Charge transfer reactions in electrochemical and chemical processes,* New York, London, Consultants Bureau, 1986, 326 p.
36. **D. E. Khoshtariya,** V. V. Topolev and L. I. Krishtalik. *Bioorganicheskaya Khimiya* Vol. 4, No. 10, 1978, 1341 p., Russian.
37. **L. I. Krishtalik**. *Khimicheskaya fizika* Vol. 6, No. 6, 1987, 836 p., Russian.

KINETICS REGULARITIES OF EMULSION COPOLYMERIZATION OF ACRYLIC MONOMERS OF VARIOUS POLARITY

S. A. Kulikov,[1] N. V. Yablokova,[1] Yu. A. Alexandrov,[1] T. V. Nikolaeva[2] and A. E. Kulikova[2]
[1]Scientific Research Institute of Chemistry, Nizhegorodsk State University
[2]V. A. Karghin Scientific Research Institute of Polymers, Dzerzhinsk City

Abstract—Emulsion (co)polymerization of butyl acrylate with methacrylic acid initiated by the persulfate ion is studied. The factors influencing the rate of the initiator degradation and the rate of copolymerization (concentrations of the emulsifier, monomers, inhibitors, temperature, etc.) are investigated. The acceleration of persulfate decomposition with increasing concentration of the emulsifier in the presence of benzoquinone and hydroquinone is found to be due to direct reaction of persulfate with substituted hydroquinone. The kinetic orders of the initiator degradation reaction with respect to all the components of the system as well as the parameters of the copolymerization reaction of butyl acrylate with methacrylic acid are determined. The effect of changes in the concentration of the emulsion system components on the mechanisms of the formation of polymeric and monomeric particles is discussed.

The negative order of the dependence of the copolymerization rate on the emulsifier concentration in the presence of inhibitors is established and this fact is discussed. A certain technique of emulsion-precipitation copolymerization of acrylic monomers in aqueous salt solutions is proposed. This technique enables the production of loosely crosslinked copolymers containing up to 80 mass% of acrylic acid units and exhibiting high viscosity characteristics.

1. INTRODUCTION

The purpose of this paper is to examine the kinetic regularities of emulsion copolymerization of two monomers distinguished by polarity. The polarity of monomers determines their hydrophilic and hydrophobic properties which, in turn, largely determine the mechanism of the process. Even in the presence of emulsifiers the hydrophilic monomers react according to the laws of polymerization in solution. Of great theoretical and practical interest is a combination of hydrophilic and hydrophobic properties within one copolymer. Such copolymers exhibit a number of valuable properties, viz., high rate of swelling in water and polar solvents, high thickening power, stability of viscosity in a wide temperature range, etc.

Moreover, the investigation of kinetic regularities and mechanism of copolymerization in monomers with various solubility of the aqueous phase contributes to the in-depth understanding of the processes occurring in emulsion polymerization (EP).

A detailed analysis of studies on EP of nonpolar and polar monomers is given in Refs. [1–4]. The idea of Smith and Ewart [5] about the amount of particles present in the emulsion process as the major kinetic factor still remains a fruitful idea in modern studies. Currently quite a number of papers are dedicated to the structure of the polymer-monomeric particles (PMP) and the place where polymerization occurs, the possibilities for the particles of complex structure to be formed [4]. The phenomena observed in EP are so different in their variety that it is impossible yet to interpret them within a single theory.

A characteristic feature of the studies on EP carried out in recent years consists in the specific conditions of the experiment which enable to simulate definite kinetic or topochemical problem. Investigations in model systems are perspective since they allow one to record separate specific parameters of EP and to fix more accurately the local zones where the elementary processes proceed.

Emulsion copolymerization of the polar monomer, methacrylic acid (MAA), and the low-polar monomer, butyl acrylate (BA), were chosen as a model system. The present paper deals with the initiation stage of polymerization, the mechanism of PMP nucleation, the effect of various factors on kinetic regularities of emulsion polymerization. A separate chapter is devoted to the synthesis of the loosely crosslinked copolymer of BA with MAA with a certain set of properties. The effect of the oxidation inhibitors on the process of emulsion copolymerization is studied.

2. EXPERIMENTAL

BA was distilled over CaH_2 under reduced pressure. MAA was purified many times by recrystallization. APS was recrystallized from water, the crystals were washed with methanol and dried in the air. HQ was twice distilled in vacuum. T_m = 169–171° C. BQ was twice recrystallized from water and ligroin. T_m = 115–116° C. The content of the basic substance in all the reagents used was not less than 98.5–99%. Water was used after double distillation, sodium laurylsulfonate (SLS) was used without additional purification. The kinetics of persulfate degradation was studied in a glass thermostated reactor with a stirrer and the constant flow of argon. The PS concentration decrease was controlled iodometrically [6]. The kinetics of emulsion copolymerization (the ratio of the monomeric phase:water-1:9) was studied by the dilatometrical method or with respect to the monomer concentration decrease controlled by the method of GLC (a Tsvet-104 chromatograph with a detector of thermal conductivity). The 100 × 0.4 cm column was packed with SE-30 (5%) on chromaton N-AW-HMDS (0.2–0.25 mm). *T* of the column was 353–373 K, with helium as the carrier-gas at a flow rate of 50 ml/min.

The SLS concentration was changed in the range (0–10) × 10^{-2} mol/l, the concentration of inhibitors was varied within (0.1–5) × 10^{-3} mol/l of the mixture. The dilatomer or the reaction vessel was previously blown through with argon, all the components of the mixture were relieved of solved oxygen and saturated with argon. The polymerization rate (v) was calculated according to the slope of the maximum rate curve plotted in coordinates of monomers conversion (a,%)—the reaction time (t, min). The size and amount of latex particles after the completion of polymerization were determined with a FEK-M device using the procedure proposed in Ref. [7].

The oxidation-reduction reaction of PS with HQ was investigated on a SF-46 spectrophotometer. The accumulation of BQ in the reaction products was controlled (λ-425 nm). Kinetic data was calculated with a computer by the method of linear regression analysis.

3. Initiation of Emulsion Polymerization of BA and MAA

The initiation of emulsion polymerization was carried out with the help of the most generally employed initiator—the persulfate ion (PS) (for example, ammonium persulfate (APS)).

It is known from the literature [8,9] that the rate of persulfate ion degradation depends on many factors, such as, the solution ionic strength, the presence of organic substrates and the surface of the phase interface. The authors who studied the PS degradation in emulsion and latexes indicate acceleration of the process as compared with the degradation in pure water [9,10–12]. This fact is determined by heterophase character of the system [12] and/or by induced degradation of PS in the presence of organic substrates, for example, emulsifier [11,13,14]. The second approach, in our opinion, is a more likely explanation of this fact.

The literature data [14,15] enable the following representation of the induced degradation of PS in emulsion:

$$S_2O_8^{2-} \xrightarrow{k_1} 2SO_4^{\cdot -} \tag{1}$$

$$S_2O_8^{2-} + E \xrightarrow{k_2} SO_4^{\cdot -} + E^{\cdot} + HSO_4^{-} \tag{2}$$

$$SO_4^{\cdot -} + H_2O \xrightarrow{k_3} {}^{\cdot}OH + HSO_4^{-} \tag{3}$$

$$SO_4^{\cdot -} + E \xrightarrow{k_4} E^{\cdot} + HO_4^{-} \tag{4}$$

$${}^{\cdot}OH + E \xrightarrow{k_5} E^{\cdot} + H_2O \tag{5}$$

$$E^{\cdot} + S_2O_8^{2-} \xrightarrow{k_6} SO_4^{\cdot -} + \text{product} \tag{6}$$

$$2SO_4^{\cdot -} \xrightarrow{k_7} \qquad (7)$$

$$2\,^{\cdot}OH \xrightarrow{k_8} \qquad (8)$$

$$2E^{\cdot} \xrightarrow{k_9} \qquad (9)$$

$$E^{\cdot} + SO_4^{\cdot -} \xrightarrow{k_{10}} \qquad (10)$$

$$E^{\cdot} = {}^{\cdot}OH \xrightarrow{k_{11}} \qquad (11)$$

$${}^{\cdot}OH + SO_4^{\cdot -} \xrightarrow{k_{12}} \qquad (12)$$

where E is the emulsifier. The scheme includes the PS monomolecular degradation (reaction 1), induced degradation (reactions 3–6). In some cases [9,14] the direct interaction of PS with the substrate is taken into account (reaction 2).

The effect of the emulsifier of emulsion polymerization exemplified by sodium laurylsulfonate (SLS) on the APS degradation is studied (Figure 1). Small additives of SLS (< 10^{-2} mole/l) increase the degradation rate, which is due to the induced degradation of PS by the radicals of the emulsifier E.

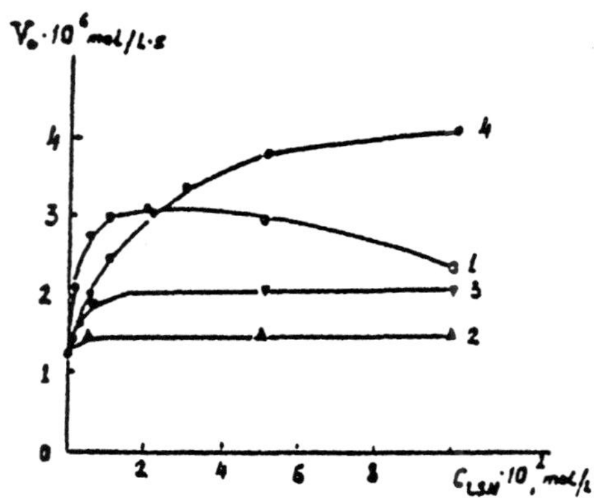

Figure 1. The effect of SLS concentration on the initial rate of APS degradation (0.5 mole/l) (curve 1) in the presence of MAA (1 mole/l) (2), BA (1 mole/l) (3) and BQ (0.005 mole/l) (4). T = 343 K.

For the critical concentration of micelle-formation (CCM) of SLS equal to 1.2 × 10^{-2} mole/l, the dependence of the initial reaction rate (v_0) on the emulsifier concentration exhibits the maximum (Figure 1, curve 1). At [SLS] > CCM, the rate of PS degradation gradually decreases, which is due to the negative salt effect of the reaction [16]. The emergence of curve 1 of dependence v_0 on the emulsifier concentration on the plateau is due to the formation of micelles. At [SLS] > CCM, the concentration of the emulsifier molecularly dissolved in water remains constant. An increase in the micelle surface with the increasing concentration of the emulsifier does not result in an increase in the rate of the process. This fact disagrees with the idea of the presence of catalysis on the surface of the phase interface for ionogenic emulsifiers of the SLS type.

Analysis of the scheme of the PS degradation process (1–12) allowed us to assume that the basic type of the breakdown is reaction 11, for the radical leading the chain is radical E˙. The reaction of E˙ with the ion-radical $SO_4^{\dot{-}}$ is hindered due to the fact that radical E˙ has the negative charge on the terminal sulfate group. Following the principle of quasistationarity and neglecting bimolecular reaction 2, we obtained the expression for the PS flow-rate in the presence of the emulsifier:

$$v = k_1 [S_2O_8^{2-}] + \left(\frac{k_1 k_5 k_6}{k_{11}} \right)^{1/2} \left[S_2O_8^{2-} \right] \cdot E^{1/2} \quad (13)$$

The orders of the process with respect to PS and the emulsifier result from equation 13: n_{APS} = 1 and n_{SLS} = 0.5, which is in good agreement with the orders determined experimentally: n_{APS} = 1.10 ± 0.05 and n_{SLS} = 0.42 ± 0.05 for the concentration of SLS < CCM. The activation energy of the APS degradation, E_a, gross process 92 ± 5 kJ/mole, log k_0 - 9.8 ÷ 0.7, which is considerably less than the values obtained for the PS degradation in pure water: E_a = 140 kJ/mole, log k_0 = 14 [9].

The presence of the monomers MAA and BA decreases the PS degradation rate in aqueous solution of the emulsifier (Figure 1, curves 2 and 3), which results from the inhibition of the induced PS degradation by the monomers [12,15]

$$SO_4^{\dot{-}} (\dot{}OH, E^{\cdot}) + M \xrightarrow{k_{14}} R^{\cdot} \quad (14)$$

In the presence of MAA, v_0 of the APS degradation is much lower than the rate of degradation in the absence of the monomer. The decrease of v_0 is due to high concentration of MAA in water and high activity of the latter in the reaction with free radicals.

In the presence of BA the dependence of the PS degradation rate, v_0, on the concentration of SLS is presented by curve 3. At [SLS] > CCM, the curve exceeds the limit at much higher values of v_0 than in the presence of MAA. BA poorly dissolves in water

Table 1. Kinetic Parameters of APS Degradation (0.01 mole/l) in Water and under the Conditions of Emulsion Copolymerization of BA (40 mole %) with MAA × C_{SLS} = 0.11 mole/l, C sub {(NH sub 4) sub 2 SO sub 4}~ = ~0.43 mole/l

	$k \times 10^5$, s^{-1}		
T,K	in water	in polym. medium	$Z=k/k_{H_2O}$
323	0.20 ± 0.02	0.94 ± 0.05	4.7
333	0.80 ± 0.01	3.5 ± 0.4	4.4
343	2.9 ± 0.1	6.1 ± 0.3	2.1
353	8.7 ± 0.3	14 ± 2	1.6

($2.3 \cdot 10^{-2}$ mole/l [1]). It mainly accumulates in micelles and a negligible amount of BA dissolved in the aqueous phase only partly suppresses reaction 6 of induced degradation of PS.

In considering the real process occurring in emulsion polymerization, it should be taken into account that the rate of PS degradation in the system under polymerization is substantially different from the rate of the process proceeding in pure water. Table 1 represents the comparison of the kinetic parameters of the process of PS degradation in pure water and in the emulsion polymerization system. The rate constants of the APS degradation in the emulsion medium exceed the values obtained in pure water by 1.6–4.7 times, with the apparent activation energy E_0 of the process decreasing up to 82 ± 8 kJ/mole and log k_0 to 8.3 ± 1.2. In addition to the accelerating effect of SLS, the pH of the medium and the ionic strength of the solution also affect the rate of PS degradation in emulsion polymerization [1].

If we take into account the mechanism of acid-catalytic degradation of PS known in the literature [9,17], the accelerating effect of the hydrogen ions in the polymerization system considered herein at 343 K can be described by equation 15

$$k = 2.2 \cdot 10^{-5} + 2.9 \cdot 10^{-3}[H^{+}] \quad (15)$$

and the effect of the ionic strength [*J*] of the solution by equation 16:

$$-1gk = 4.14 + 0.26[J] \quad (16)$$

For practical purposes it is important to specify the effect of the inhibitors of the monomer oxidation on the PS degradation, for example, such inhibitors as widely used compounds of quinoid structure. Experimental data on the PS degradation in the presence of benzoquinone (BQ) given below allow us to supplement scheme 1_12 of the PS degradation with the following reactions:

$$E^{\cdot} + 0 = \quad = 0 \xrightarrow{k_{17.}} HO - \overset{E}{\quad} - 0^{\cdot} \xrightarrow{HOH} HO - \overset{E}{\quad} - OH \quad (17)$$

$$S_2O_8^{2-} + HO - \underset{E}{\quad} - OH \xrightarrow{k_{18}} SO_4^{\cdot -} + HSO_4^{-} + {}^{\cdot}O - \underset{E}{\quad} - OH \xrightarrow{PS} O = \underset{E}{\quad} = O$$

(18)

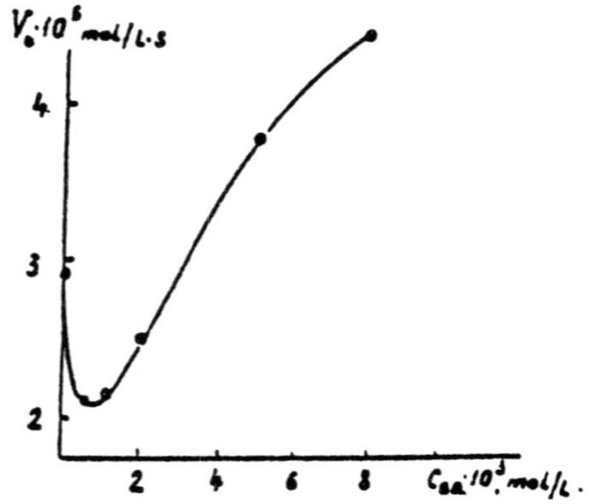

Figure 2. Dependence of the initial rate of APS degradation (0.05 mole/l) on the BQ concentration in water in the presence of 0.05 mole/l of SLS. *T* = 343 K.

The presence of BQ in the system changes the kinetic regularities of the PS degradation in the emulsifier solution. The dependence of v_0 on the BQ concentration is of a complicated nature (Figure 2). At the beginning the PS degradation rate decreases, passes through the minimum and then increases. At low concentrations, BQ reacts as inhibitor suppressing the free-radical chain degradation due to reaction 17, which results in a decrease of the rate of the gross process. The acceleration of the PS degradation observed at $C_{BQ} > 5 \times 10^{-4}$ mole/l is determined by the course of reaction 18. This reaction is similar to the oxidation-reduction reaction of PS with hydro-quinone (HQ) known in the literature [18,19]:

$$\begin{aligned} 2S_2O_8^{2-} + HO- \quad OH \longrightarrow 2SO_4^{\cdot-} \\ + 2HSO_4^{-} + O= \quad =O \end{aligned} \tag{19}$$

Reaction 19 was investigated by the spectrophotometric method, the results of kinetic measurement are given in Table 2. The activation energy of the HQ oxidation by persulfate is equal to 72.7 ± 0.6 kJ/mole. The reaction proceeds at a high rate. For example, at 343 K, HQ quantitatively changes to BQ within 14 minutes. For this reason, after this period of time the effect of HQ on the PS degradation process becomes completely similar to the effect of BQ.

Table 2. Kinetic Parameters of the Oxidation-reduction Reaction of APS with HQ. $C_{HQ} = 5 \times 10^{-3}$ mole/l, $C_{APS} = 5 \times 10^{-2}$ mole/l

T,k	$v_{0 \times 106}$, mole/l s	10 $t_{1/2}$, min
295	0.3	500
320	3.4	49
330	7.3	28
340	17.3	16
343	21.5	14

If HQ readily changes to BQ, the latter, in its turn, changes to the HQ analog by adding the emulsifier radical, $E^{\times}$, the HQ analog exhibiting the surface active properties. Thus, in the presence of the emulsifier and BQ(HQ) the following chain of successive transformation takes place:

$$HO- \quad -OH \xrightarrow{PS} O= \quad =O \xrightarrow{E^{\cdot}} HO- \quad ^{E} \; -OH \xrightarrow{PS} O= \quad ^{E} \; =O \longrightarrow ect \qquad (20)$$

The flow-rate of BQ according to reactions 17 and 18 was experimentally measured by the method of gas-liquid chromatography. At 343 K, the BQ concentration in the solution containing 0.05 mole/l of APS, 0.1 mole/l of SLS and 0.05 mole/l of BQ decreases to 0.018 mole/l, i.e., more than one half.

In the presence of BQ, the curve of the dependence of v_0 of the PS degradation on the SLS concentration is not seen to twist or emerge on the plateau (curve 4, Figure 1). The fact that at C_{SLS} > CCM the curve does not exceed the limit in the presence of BQ can be explained as follows. BQ concentrates in the emulsifier micelles due to poor solubility in water. The $^{\times}OH$ radicals penetrating the micelles (the ion-radicals $SO_4^{\cdot-}$ do not enter into the micelles of the ionogenic emulsifier because of electrostatic repulsion) give rise to radicals $E^{\times}$; attacking BQ in micelles the latter contribute to the formation of substituted HQ with the surface-active substitute. Due to continuous exchange of the emulsifier between the solution and the micelle surface the product of reaction [17] passes to the aqueous phase where it reacts with PS according to reaction 18. When the SLS concentration exceeds CCM, the concentration of the substituted HQ continuously increases and, therefore, v_0 increases without emerging on the plateau.

Proceeding from the above consideration we can obtain the following expression for the PS flow-rate velocity in the presence of SLS and BQ.

$$V = 1/2k_1[S_2O_8^{2-} + \left\{1/2k_1[S_2O_8^{2-}]^2 + \frac{K_1k_5k_{17}}{k_{11}}[S_2O_8^{2-}][E][BQ]\right\}^{1/2} \qquad (21)$$

The orders we observed in the degradation reaction of PS throughout all the reagents were equal to: $n_{APS} = 0.95 \pm 0.06$, $n_{SLS} = 0.24 \pm 0.02$, $n_{BQ} = 0.27 \pm 0.04$. These orders agree with equation 21 if we take into account that the contribution of the latter member in the total rate of the process is not large. The activation energy of the PS degradation in the presence of BQ was equal to 84.9 ± 1.7 kJ/mole, log $k_0 = 8.9 \pm 0.1$. The values obtained for E_a and log k_0 are even much less than in the absence of

BQ and these parameters are much more different from the PS degradation in pure water.

The effect of the monomers MAA and BA on the rate of PS degradation in the presence of SLS and BQ is described by the same curves 2 and 3 (Figure 1) as in the case where BQ was absent. Consequently, irrespective of their solubility in water, the monomers suppress reaction 18.

The efficiency of initiation by persulfate in aqueous solutions is close to unity [20]. It is difficult to speak about the efficiency of the initiator in the heterophase polymerization. In emulsion polymerization of monomers with various polarity the reaction of initiation proceeds in the aqueous phase. If we assume that the growing polymer radicals do not recombine with the original radicals, then the efficiency of initiating the emulsion polymerization can be identified with the efficiency of polymerization initiation in the aqueous phase. We simulated the conditions for PS degradation in the aqueous phase upon emulsion polymerization of BA with MAA (1:1) and the ratio of the monomer phase:water—1:9. The coefficient of distribution being taken into account, the MAA concentration in the aqueous phase equals to $8 \cdot 10^{-2}$ mole/l. At 333 K and the SLS concentration equal to $1.8 \cdot 10^{-2}$ mole/l, the efficiency of the PS initiation ($2.2 \cdot 10^{-3}$ mole/l) determined by the inhibition method (4-oxo-2,2′,6,6′-tetramethyl-piperidineoxyl-1) was found to be equal to 0.51 ± 0.03. In the absence of the emulsifier the initiation efficiency is higher, 0.68 ± 0.04.

4. Formation of PMP in the Conditions of Emulsion Polymerization of BA with MAA

The mechanism of the formation of PMP is a key problem in the study of EP.

Three probable mechanisms of the PMP nucleation are considered in the literature [1,4,21]: (1) the micellar mechanism; (2) homogeneous type of nucleation; (3) formation of PMP from the monomer droplets. Also exists an intermediate type of nucleation when oligomeric radicals produced in the aqueous phase do not have enough time to form a separate phase and are captured by the micelles and PMP present there.

The third type of particle formation in the conditions of copolymerization of BA with MAA is unlikely to exist. The conditions for this type of PMP formation to be noticeable are attained only due to a strong hydrodynamic impact on the emulsion system leading to the separation of monomeric droplets and an increase in their total surface. As a result, the active radicals from the aqueous phase can more readily enter into the droplets. The rate of stirring in our experiments (not more than 400 r/min) is not sufficient for this purpose.

To evaluate the contribution of each of the above-mentioned types of PMP nucleation in the process of the formation of latex copolymer of BA with MAA we investigated the effect of the emulsifier concentration (Figure 3), the initiator and the ratio of concentrations of BA and MAA (Figure 4) on the kinetic of (co)polymerization.

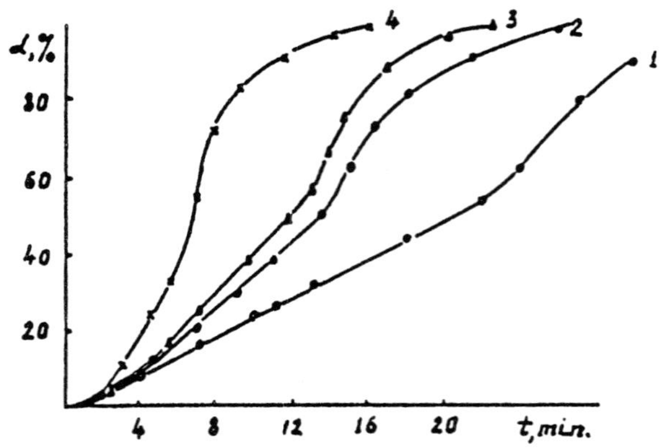

Figure 3. Copolymerization of BA (30 mole %) with MAA · $C_{APS} = 2.2 \cdot 10^{-3}$, $C_{AS} = 2.3 \cdot 10^{-2}$, $C_{SLS} = 0(1), 4.5(2), 8.9(3)\ 30 \cdot 10^{-3}$ mole/l (4). Monomeric phase:water-1:9. $T = 333$ K.

In the absence of the emulsifier (curve 1, Figure 3) the origination of PMP follows the homogeneous mechanism. The initial particles are stabilized by the polar groups of MAA and the initiator. BA enters into the copolymerization process at the conversion of about 50%. As a result, a mixture of the copolymer of BA and MAA and the homopolymer MAA is formed. As the emulsifier concentration increases, BA enters into the process of copolymerization at much earlier stages, with the amount of the homopolymer MAA in the final product decreasing (curve 2–4, Figure 3). At the concentration of SLS > CCM, nucleation follows the intermediate type. The rectilinear initial section of the curve corresponding to the dissolving polymerization of MAA disappears.

An increase in the concentration of the initiator results in a higher rate of initiation and shortening of the initial section of the curves corresponding to homopolymerization of MAA. Increasing rate of initiation increases the number of oligomeric radicals of MAA in the aqueous phase, which makes it difficult for oligomeric radicals to reach the critical molecular mass in order to form a separate phase. Thus, an increase of the initiator concentration results in a transition from the homogeneous type of nucleation to the intermediate one and then, in case of a sufficient amount of the emulsifier, to the micellar mechanism.

If the SLS concentration is less than CCM, the process of polymerization of pure BA (curve 1, Figure 4) exhibits the "induction period" determined by the difficulties in the formation of

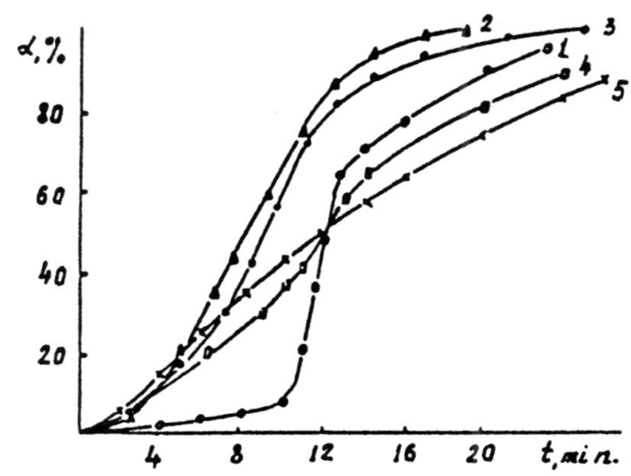

Figure 4. Polymerization and copolymerization of BA and MAA at 333 K. $C_{SLS} = 1 \cdot 10^{-2}$, $C_{AS} = 2.3 \cdot 10^{-2}$, $C_{APS} = 2 \cdot 10^{-3}$ mole/l. 1–100 mole % BA, 2–60 mole % BA, 3–40 mole % BA, 4–20 mole % BA, 5–100 mole % MAA. Monomeric phase: water-1:9.

PMP due to the absence of micelles. Polymerization of BA is characterized by the presence of the gel-effect at the early stages [21]. For this reason, the sections of the maximum rate can be identified with the participation of BA in the process of copolymerization since the gel-effect is not typical for the polymerization of MAA (curve 5, Figure 4). At low concentrations of BA, the polymerization of MAA in water and BA in the formed PMP is observed in the system (Curve 4, Figure 4). When the BA concentration exceeds 30 mol. %, the copolymerization of BA with MAA proceeds according to the intermediate type (curves 2 and 3, Figure 4).

Some salting-out agents are used in practical work to decrease the concentration of the polar monomer in water and to change it to the monomer phase. Figure 5 represents the effect of the salting-out agent (ammonium sulfate, AS) on the kinetics of copolymerization of BA with MAA. In the presence of AS the copolymerization rate becomes lower, because the monomers are expelled from the aqueous phase and the rate of reaction 14 decreases. The coefficient of MAA distribution between water and the monomer phase after incorporation of AS (0.21 mole/l) changes from 0.16 to 0.08. In this case, pH of the medium increases from 3.5 to 4.5. In addition to decreasing the total rate of the process, the salting-out agent also decreases the fraction of the homopolymer of MAA. At the concentration of the emulsifier significantly exceeding CCM, the mechanism of PMP origination with increasing concentration of AS approaches the micellar mechanism.

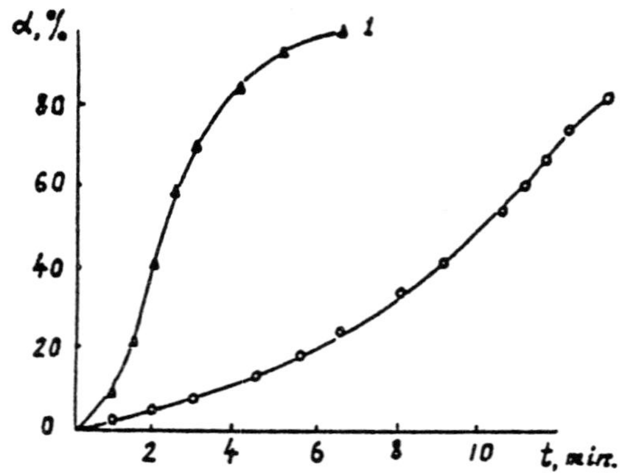

Figure 5. The effect of the salting-out agent $(NH_4)_2SO_4$ on copolymerization of BA (40 mole %) with MAA at 343 K. $C_{APS} = 1 \cdot 10^{-2}$, $C_{SLS} = 0.1$. $C_{AS} = 0(1)$ and 0.23 mole/l (2). Monomeric phase:water-1:9.

Thus the micellar mechanism of nucleation is realized both for the emulsion polymerization of BA and for the copolymerization of BA with MAA when the fraction of MAA is less than 30 mol. % at high concentrations of the emulsifier and the salting-out agent. The intermediate type of nucleation is realized at the MAA concentrations from 30 to 70 mol. % in a mixture with BA and the concentration of the emulsifier higher than CCM. The homogeneous nucleation is realized at the concentrations of the emulsifier below CCM as well as at the high concentration of MAA in a mixture exceeding 70 mol. %.

5. Kinetic Parameters of Emulsion Copolymerization of BA with MAA

Realization of the homogeneous and intermediate types of PMP nucleation results in deviation of kinetic and activation parameters of copolymerization of monomers with different polarity from the law described by the theory of Smith and Ewart [5]. The order of the copolymerization rate with respect to the initiator was found to be equal to 0.25 ± 0.01, with respect to the emulsifier - 0.40 ± 0.05 (Figure 3). In the absence of the salting-out agent when the contribution of the MAA homopolymerization to the total rate of the process is great the order with respect to the emulsifier decreases to 0.23 ± 0.01 [22].

The relative effective activities of BA (r_1) and MAA (r_2) in emulsion copolymerization at 343 K were determined at various conversion of monomers. At low conversions (7–10%), $r_1 = 0.24 \pm 0.23$, $r_2 = 1.27 \pm 0.07$. At the initial stage of copolymerization, the copolymer is enriched with units of MAA. Analysis of the copolymer at deep conversion degrees (~ 80%) has shown that the copolymer composition is closer to the initial ratio of monomers and is less enriched with MAA: $r_1 = 0.5 \pm 0.15$, $r_2 = 1.1 \pm 0.2$.

Activation parameters of the process of copolymerization slightly depend on the change in the ratio of the mixture components and is equal to the values of $E_a = 39 \pm 4$ kJ/mole and log $v_0 = 5.8 \pm 0.8$. However, these parameters decrease as the salting-out agent is introduced. In the presence of 0.023–0.23 mole/l of AS the activation energy of the process equals 25–27 kJ/mole and log $v_0 = 4.4 - 3.1$.

6. Preparation and Properties of Loosely Crosslinked Copolymer of BA with MAA

Copolymers with high content of the polar monomer are not much feasible for practical use because of low viscosity of aqueous and water-alkali solutions. A small amount of a crosslinking agent (0.5–2.5

mass. % of the monomers mass) is added to the polymerizing system to increase the viscosity of these solutions. This small amount of crosslinking agents provides about one crosslinking per 100–200 monomeric units. The viscosity of the 2.5% solution of the copolymer of BA with MAA containing 64–86 mol. % of MAA (the solution is neutralized by ammonia) exceeds 4000 Pa.s (Table 3).

A new method was developed for obtaining latex copolymer in order to produce copolymers with this high content of MAA. The method combines emulsion and dissolving-precipitating polymerization.

As shown above, the salting-out agent incorporated in the system at the stage of synthesis enables the formation of the copolymer with a higher content of the polar monomer. Table 3 shows that the copolymer containing 64 mol. % of MAA cannot be isolated by ordinary emulsion polymerization. We managed to isolate a copolymer containing 86 mol. % of MAA with the use of the emulsifier and the salting-out agent in great content. The copolymer of BA with MAA does not form when the initial monomeric mixture contains less than 86 mole % of MAA. The monomers are polymerized independently of each other, which was noted above in the analysis of kinetic curves of copolymerization of BA with MAA. If the MAA content in the mixture is less than 60 mole %, the loosely crosslinked copolymer is not practically soluble in water solution of ammonia. Thus the high viscosity of solutions is provided by the copolymers with a narrow range of the monomers relations (64–84 mole % of MAA).

Table 3. Dependence of the Viscosity of 2.5%-ammonia solution of the Copolymer of BA with MAA on the Copolymer Composition. The Crosslinking agent-Dimethylacryl Ester of Ethyleneglycol (1.5 mass. %). The Salting-out agent—$(NH_4)_2SO_4$—5 mass %.

MAA Content in the Initial Mixture of Monomers, mole %	Viscosity according to Brookfield, Pa.s
50	copolymer does not dissolve
64*	copolymer is not isolated
64	50
73	10
78	5
86	4
93	copolymer is not formed

*Synthesis proceeds without a salting-out agent.

It follows from Figure 4 that copolymerization of MAA with BA proceeds fast and uniformly if the MAA content in the initial mixture is equal to 30–70 mole %. Hence, the optimal concentration of MAA in the initial mixture should be 64–70 mole % in order to produce loosely crosslinked copolymers with good viscosity parameters.

7. THE EFFECT OF INHIBITORS ON EMULSION COPOLYMERIZATION OF BA WITH MAA

Practically no information about the behavior of inhibitors in emulsion systems is presented in the literature [23,24]. Of great interest is to know the inhibitors transformation in the real systems since the purification of the monomers from the inhibitors of oxidation is a stage which makes the process of polymerization more expensive.

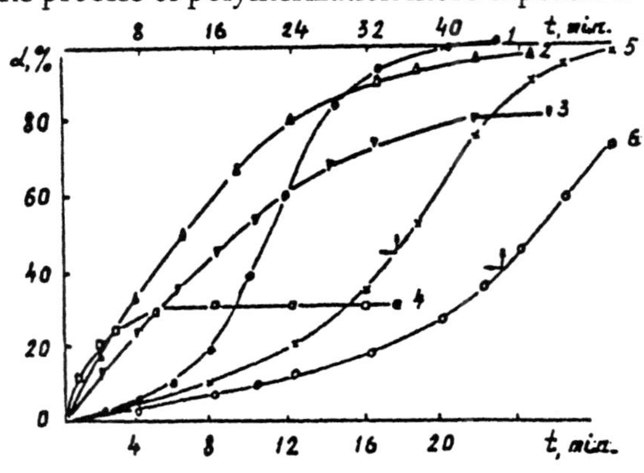

Figure 6. The effect of HQ and BQ on copolymerization of BA (50 mole %) with MAA. $C_{APS} = 2.2 \cdot 10^{-3}$, $C_{SLS} = 1.5 \cdot 10^{-2}$ mole/l, $C_{HQ} = 0(1)$, 0.6 (2), 1.1(3) and $2.2 \cdot 10^{-3}$ mole/l (4). $C_{BQ} = 2.2$ (5) and $4.4 \cdot 10^{-4}$ mole/l (6). $T = 333$ K. Monomeric phase:water-1:9.

As noted above, PS reacts with HQ through the redox reaction (reaction 18 (19)) which results in a significant increase in the yield of free radicals in the system. Figure 6 shows the curves of copolymerization of BA with MAA in the absence of inhibitors (curve 1) and in the presence of HQ (2–4) and BQ (5,6). An increase in the initiation rate in the presence of HQ increases the initial rate of copolymerization, the s-shaped character of curve 1 disappears and copolymerization starts straight at the maximum rate (curves 2 and 3, Figure 6). At the ratio [HQ]/[PS] > 0.5, the initiator is completely used up within a short period of time, and it is not necessary for the monomers to attain a 100% conversion (curve 4).

The mechanism of the PMP origination changes in the presence of HQ. The effect due to the presence of HQ in the system is similar to the increasing concentration of the initiator, which results in the less reaction time for copolymerization in the water phase. Small additions of HQ enable the reaction of copolymerization to pass over to micelles much faster, which decreases the fraction of the homopolymer of MAA in the product and improves the structural uniformity of the copolymer.

The additions of BQ (curves 5 and 6, Figure 6) increase the time during which the curve of copolymerization attains the maximum rate and decrease the rate of the process. BQ does not suppress the MAA polymerization in the water phase completely but only decreases its rate. The mechanism of the PMP origination in the presence of BQ remains the same, which is verified by the retained s-shape of curves 5 and 6 of copolymerization. However, the time of the development of MAA polymerization becomes longer in the water phase, the amount of the homopolymer of MAA increases resulting in deterioration of the structural uniformity of the copolymer of BA with MAA.

The negative order (n) of the dependence of the (co)polymerization rate on the concentration of the emulsifier in the presence of HQ and BQ is determined (Table 4). A similar fact is previously described in Ref. [23]. The inverse order with respect to the emulsifier exhibits itself only for BA and the mixture of monomers containing BA, i.e., for the nonpolar monomer polymerized according to the micellar mechanism. For MAA polymerized in the aqueous phase this situation is not observed (Table 4).

It has been found that in the presence of HQ or BQ the number of particles N in the system does not increase with an increase of the emulsifier concentration, as it is required by the theory [5], but, on the contrary, it decreases (Table 5). The presence of the inhibitor in the emulsion sharply diminishes the number of latex particles in the final stage of the process. As the concentration of the emulsifier increases, the amount of the potential centers of copolymerization increases.

Table 4. Effect of the Emulsifier Concentration on the Parameters of (co)polymerization of BA and MAA. *T* 333 K = 2.2×10^{-3} mol/l, $C_{HQ} = C_{BQ} = 2.2 \times 10^{-4}$ mol/l

$C_{SLS} \times 10^2$ mol/l	$v_p \times 10^2$, % s^{-1} BA without addition	BA in the presence of HQ	BA in the presence of BQ	MAA without addition	MAA in the presence of HQ	MAA in the presence of BQ	BA ± MAA (1:1) without addition	BA ± MAA (1:1) in the presence of HQ	BA ± MAA (1:1) in the presence of BQ
0	-	-	-	6.1	7.5	3.7	10.2	-	-
0.46	17.5	19.5	4.4	-	-	3.3	19.2	12.0	11.4
0.91	20.6	15.6	4.0	6.5	10.2	2.8	24.1	11.6	8.6
1.82	24.2	12.9	3.7	6.6	11.8	3.3	28.8	9.4	8.2
3.64	30.2	10.8	3.2	6.6	13.5	3.6	33.6	5.8	7.6
5.50	-	-	-	-	-	4.0	-	-	7.2
r	0.3	-0.3	-0.16	0	0.2	0.2	0.25	-0.31	-0.14

The fraction of these centers decreases due to reaction 17 in the course of which was formed the substituted semiquinone radical exhibiting high activity with respect to radicals E˙. An inhibitor having surface-active properties forms on the interface. This inhibitor acts as a trap of free radicals thus preventing their penetration into the micelles and PMP from the water phase.

Table 5. Dependence of the Number of Latex Particles of the Polymer of BA and Copolymer of BA with MAA on the Concentration of the Emulsifier. T = 333 K. $C_{APS} = 2.2 \times 10^{-3}$ mol/l, $C_{HQ(BQ)} = 2.2 \times 10^{-4}$ mol/l.

	$N \times 10^{-11}$ cm^{-3}					
	BA			BA ± MAA (1:1)		
$C_{SLS} \times 10_2$ mole/l	without inhibitor	HQ	BQ	without inhibitor	HQ	BQ
0.46	41	9.8	1.9	17	3.8	10.9
0.91	71	7.3	1.7	30	2.0	5.1
1.82	138	6.1	1.4	26	1.3	1.4
3.64	460	5.1	1.2	41	1.5	0.92

These effects are not typical for MAA polymerized in solution.

Thus it follows that with an increase in the concentration of the emulsifier the penetration of free radicals into monomers becomes constrained and the number of "dead" particles which do not participate in polymerization increases and the amount of latex particles N diminishes.

8. CONCLUSIONS

The present paper contributes to the understanding of the mechanism and the kinetic regularities of emulsion (co)polymerization of monomers with various polarity. The factors affecting the rate and efficiency of initiation upon polymerization in emulsion have been studied. It was found that the emulsifier increases the rate of degradation though it decreases the effectiveness of initiating the persulfate ions. Monomers suppress the induced degradation of the initiator. The rate of inhibition of this process depends on the activity of monomers and their solubility in the water phase. The inhibitors of oxidation of monomers HQ and BQ increase the rate of degradation of persulfates.

The effect of the concentrations of the components of emulsion (co)polymerization as well as of the addition of the salting-out agent and the inhibitor on the mechanism of PMP formation upon emulsion copolymerization of BA with MAA was identified. The kinetic and energetic parameters of emulsion polymerization were calculated. It has been shown that small additions of HQ have a favorable effect on the kinetics of emulsion copolymerization and the properties of products. The presence of BQ causes deterioration of these indicators. A certain interpretation of the inverse order of the (co)polymerization rate with respect to the emulsifier observed in the presence of inhibitors was proposed. Also proposed is a method of producing the copolymer with a greater amount of units of the polar monomer.

9. References

1. **V. I. Eliseeva, S. S. Ivanchev, S. I. Kuchanov,** and **A. A. Lebedev**. *Emulsionnaya polimerizatsiya i ee primenenie v promyshlennosti* (Emulsion Polymerization and its Practical Use in Industry). Moscow, Khimiya, 1976, 240 p., Russian.
2. **U. Cooper**. Emulsion Polymerization, in *Reactivity, Mechanism and Structure in Polymer Chemistry* (A. D. Jenkins and A. Ledwith, eds.), J. Wiley and Sons, Ltd., 1974, 276 p.
3. **V. N. Pavlyuchenko,** and **S. S. Ivanchev**. *Uspekhi khimii,* Vol. 50, No. 4, 1981, 715 p, Russian.
4. **Z. Song,** and **G. Poelein**. *J. Makromol. Sci. Chem.,* Vol. A25, No. 4, 1988, 403 p.
5. **W. V. Smith,** and **R. W. Ewart**. *J. Chem. Phys.,* Vol. 16, 1948, 592 p.
6. **I. M. Kolthoff,** and **E. M. Carr**. *Analyt. Chem.* Vol. 25, No. 2, 1953, 298 p.
7. **P. L. Slonim**. *Optika i spektroskopiya* (Optics and Spectroscopy), Vol. 8, No. 1, 1960, 98 p. Russian.
8. **I. M. Kolthoff, E. Meechan,** and **E. Carr**. *J. Am. Chem. Soc.,* Vol. 75, No. 4, 1953, 1439 p.
9. **I. M. Kolthoff,** and **I. Miller,** *J. Am. Chem. Soc.,* Vol. 73, No. 8, 1951, 3055 p.
10. **B. W. Brooks,** and **B. O. Makajola**. *Makromol. Chem.* Rapid Com. Bd. Vol. 2, No. 1, 1981, 69 s.

11. **M. S. Ryabova, S. N. Sautin,** and **N. N. Smirnov**. *Zh. prikl. khimii.,* Vol. 50, No. 8, 1977, 1719 p., Russian.
12. **M. S. Ryabova, S. N. Sautin,** and **N. N. Smirnov**. *Zh. prikl. khimii,* Vol. 51, No. 9, 1978, 2059, Russian.
13. **S. Morris** and **A.G. Parts**. *Makromol. Chem.,* Bd. 119, 1968, S. 212.
14. **W. K. Wilmarth,** and **A. Hain**. *Mechanism of Oxidation by Peroxydisulfate.* Peroxide Reaction Mechanism (J. O. Edward, ed.), New York Interscience Publ., 1962, 175 p.
15. **A. A. Berlin,** and **V. N. Kislenko**. *Zh. obshch. khimii,* Vol. 59, No. 1, 1989, 3 p., Russian.
16. **N. V. Yablokova, S. A. Kulikov, T. R. Karpinskaya,** and **Yu. A. Alexandrov**. *Radikalnye reaktsii metallorganicheskikh soedinenii* (Radical Reaction of Organometal Compounds), Gorky, Gorky University, 1986, 79 p., Russian.
17. **U. Chandra Singh,** and **K. Vencatarao**. *Int. J. Chem. Kinet.,* Vol. 13, No. 6, 1981, 555 p.
18. **R. C. R. Bacon**. *Trans. Far. Soc.,* Vol. 42, 1946, 140 p.
19. **B. A. Dolgoplosk,** and **T. I. Tinyakova**. *Generirovanie svobodnykh radikalov i ikh reaktsii* (Generation of Free Radicals and their Reactions), Moscow: Nauka, 1982, 92 p., Russian.
20. **E. N. Zil'berman, L. I. Abramova,** and **T. A. Nekrasova**. *Vysokomol. soed.,* Seriya A., Vol. 20, No. 10, 1978, 2331 p., Russian.
21. **V. I. Eliseeva,** and **N. P. Zhukov**. *Vysokomol. soed.,* Seriya A., Vol. 23, No. 7, 1981, 1540 p., Russian.
22. **S. A. Kulikov, N. V. Yablokova, T. V. Nikolaeva,** and **Yu. A. Alexandrov**. *Vysokomol. soed.,* Seriya A., Vol. 31, No. 11, 1989, 2322 p., Russian.
23. **J. Capek,** and **J. Bartoh**. *Makromol. Chem.,* Vol. 186, No. 6, 1985, 1297 p.
24. **J. Bartoh,** and **V. Juranicova**. *Makromol. Chem.,* Vol. 190, No. 4, 1989, 763 p.

INHIBITING ACTION OF COPPER(2^+) AND COBALT(2^+) CHELATES IN OXIDATION PROCESS OF POLYETHYLENE GLYCOLS IN SOLUTION

A.B. Mazaletskii and V.G. Vinogradova
Institute of Chemical Physics,
Russian Academy of Sciences,
Moscow, Russia

ABSTRACT—Antioxidizing properties of some chelate compounds of copper(2^+) and cobalt(2^+) with nitrogen-, oxygen- and sulfur-containing chelate active centers in radical-chain oxidation of triethylene glycol (TEG), polyethylene glycols of various molecular masses (PEG-400, PEG-1500, PEG-1400) and dimethyl ether of polyethylene glycol (DMEP-5000) in solution by molecular oxygen has been investigated. Anti=radical action of these complexes has been studied upon initiated oxidation (75° C, solution in chlorobenzene); gross-inhibiting action has been examined upon autoxidation of PEG-1500 (95° C, solution in o-dichlorobenzene). All the chelates studied exhibit the inhibiting action in the course of initiated oxidation of TEG, PEG-400 and PEG-1500, with the most active compounds in the concentration 10^{-4} to 10^{-3} mol/1 completely suppressing the chain oxidation. In the process of TEG oxidation the stoichiometric coefficients f_t amount to 7.5, while in transition to PEG-400 and PEG-1500 the values of f_t decrease to 1 to 3. It results from the declining role of a-oxyperoxide ("alcoholic") radicals, which possess dual oxidation-reduction properties, in the gross-oxidation of PEG. Upon further transition to PEG-14000 and DMEP-5000 the chelates exactly fall into two groups one of which retains the inhibiting activity and the other completely loses it.

With the use of the electrochemical technique, the single-electron redox characteristics of the complexes under investigation have been identified and the correlations between these characteristics and the inhibition constants for the chelates in the oxidation of PEG-1500 have been established thus allowing the mechanism of action of metal-containing inhibitors to be specified.

Transition from the catalytic function to the inhibiting one is observed in the autoxidation of PEG-1500 upon increasing initial concentration of chelates C_0. The series of nitrogen- and sulfur-containing chelates is characterized by low values of "critical concentration" C_{cr} and high inhibiting at $C_0 > C_{cr}$.

1. INTRODUCTION

Polyethylene glycol ($HO(CH_2CH_2O)_nH$ with various degrees of polymerization as well as its derivatives are used as complex-forming agents, catalysts of "phase transfer," stationary phase in gas chromatography, plastisizers of polymeric materials, etc. [1–3]. When exposed to atmospheric oxygen, polyethylene glycol (PEG) having heterochain molecules with activated C—H bonds can undergo radical-chain oxidation in the state of melts or solution [4,5]. Kinetic regularities of PEG oxidation in inert solvents as well as the regularities of liquid-phase oxidation of low oligomers of ethylene glycol correlate with the standard scheme of chain liquid-phase oxidation of organic substances [6,7].

Simple salts and chelate compounds of variable valence metals which are single-electron redox agents actively inhibit the chain oxidation of some oxygen-containing organic substances (primary and secondary aliphatic alcohols, partially etherified polyols, etc.) which are oxidized due to the presence of α-oxyperoxyl ("alcoholic") radicals RO_2 exhibiting both the oxidizing and reducing properties [8]. In this case many times repeated chain termination is realized, which is based on a set of successive redox stages:

$$M^{n+} + RO_2 \rightarrow M^{(n+1)} + RO_2^- \quad (1)$$

$$M^{(n+1)+} + RO_2 \rightarrow M^{n+} + \text{molecular products}, \quad (2)$$

where M is a variable valence metal.

Table 1. Half-wave potentials of electrochemical oxidation ($E_{1/2\ ox.}$) and reduction of chelates I-XXVII ($E_{1/2\ r.}$) on the rotating platinum electrode relative to saturated camel reference electrode (solvent: acetonitrile-chlorobenzene¾1:1).

Chelate	$E_{1/2\ ox}$,B	$E_{1/2\ r.}$,B	Chelate	$E_{1/2\ ox.}$,B	$E_{1/2\ r.}$, B
I	0.72	-0.92	XY	1.09	-1.20
II	0.97	-0.72	XYI	1.08	-1.00
III	1.07	-0.40	XYII	1.07	-0.90
IY	0.76	-0.62	XYIII	1.10	-1.15
Y	0.67	-0.64	XIX	0.95	-0.91
YI	0.84	-0.55	XX	0.85	-0.73
YII	0.96	-0.40	XXI	1.23	-1.00
YIII	1.00		XXII	0.74	-0.29
IX	0.56		XXIII	1.30	-0.55
X	0.91		XXIY	0.72	-1.08
XI	1.28		XXY	0.62	
XII	1.38				
XIII	0.74	-0.55	XXYI	1.01	
XIY	1.10	-0.30	XXYII	0.83	

The process of PEG oxidation involves both "alcohol" peroxyradicals and "ether" peroxyradicals. The latter are formed upon splitting off the hydrogen atom from the internal CH_2 groups of the PEG molecule and do not exhibit the reducing properties. Oxidation of the lower oligomers of ethylene glycol is inhibited by salts of variable valence metals [9]. Since the role of the reactions of "alcohol" peroxyradicals should, in principle, diminish with an increase of the PEG polymerization degree, a possibility of effective stabilization of high-molecular PEG by the compounds of variable valence metals requires a special investigation.

2. Results and Discussion

We have investigated bi- and tricyclic nitrogen-, oxygen- and sulfur-containing chelates of copper(2^+) and cobalt(2^+) I-XXV as potential oxidation inhibitors. The inhibiting action of Zn(2^+) and Bi(3^+) diethyldithiocarbamates XXVI and XXII was studied for comparison purposes. They exhibit antioxidizing properties in liquid-phase oxidation of hydrocarbons whose peroxyradicals possess only oxidizing properties [10].

As shown previously, in a number of cases the antiradical activity of metal complexes correlates with their single-electron redox characteristics [10,11]. We determined the values of half-wave potentials of electrochemical oxidation ($E_{1/2\ ox.}$) and reduction ($E_{1/2\ r.}$) of chelates on the rotating platinum electrode in the "chlorobenzene-acetonitrile (1:1) "mixture in the presence of tetraethylammonium perchlorate (0.1 mol/l). The values of the half-wave potentials referenced to the saturated calomel electrode are given in Table 1.

2.1 Non-Inhibited Oxidation of Polyethylene Glycols

We studied low-temperature initiated oxidation of triethylene glycol (TEG), oligomeric PEG-400 (the average degree of polymerization $n_{av.} \simeq 9$), PEG-1500 ($n_{av.} \simeq 33$) and also high-molecular PEG-1400 ($n_{av.} \simeq 320$) and dimethyl ether of polyethylene glycol DMEP-5000 ($n_{av.} \simeq 110$). Autoxidation of PEG-1500 in solution was also investigated; a number of experiments of DMEP-5000 autoxidation were carried out.

Initiated oxidation was performed in chlorobenzene solution at 75° C, the concentration of the substance under oxidation (per 1 unit of polymeric chain) was equal to 15 mol/l (TEG), 10 mol/l (PEG-400 and PEG-1500), 6.7 mol/l (DMEP-5000) and 5 mol/l (PEG-14000). The process was carried out at the atmospheric pressure of the oxidizing gas mixture containing 20 to 100% of oxygen by volume; vigorous stirring of the solution with a magnetic stirrer provided kinetic conditions for oxygen absorption by the solution, the rate of oxygen absorption being measured on a volumetric device equipped with an automatic pressure control. Oxidation was initiated by azobisisobutyronitrile (AIBN); the initiation rate W_i was measured by the inhibitor technique with the use of chelate XXIi whose stoichiometric inhibition coefficient f_t is equal to 2 upon oxidation of hydrocarbons and cyclohexanol.

The initiation constant for AIBN (k_i) can considerably decrease with an increase in the medium viscosity due to less possibility for radicals to leave the "cage" of the solvent (decrease in the initiation efficiency). In fact, the k_i values in solutions of PEG-400, PEG-1500 and TEG equal to (7.2 to 7.5) $\cdot 10^{-5}$ s^{-1} account for 70% of the k_i value in chlorobenzene. However further increase in the medium macroviscosity upon transition to solutions of high-molecular DMEP-5000 and PEG-14000 leads to somewhat unexpected increase in k_i (up to 9×10^{-5} s^{-1} and $1.0 \cdot 10^{-4}$ s^{-1}, respectively). Probably it results from structurization of high-molecular PEG solutions and "ousting" the initiator into the volume occupied by the solvent.

The relationship $W_{ox.} \sim (O_2)^0 \cdot W_i^{1/2}$ is observed for the oxidation rate of PEG in solution. This relationship is typical for chain oxidation of organic substances with quadratic chain termination on peroxyradicals RO_2. The values of the oxidation parameter $k_2 \cdot k_6^{1/2}$ (where k_2 is the rate constant of chain propagation by peroxiradicals, and k_6 is the rate constant of quadratic chain termination) calculated per a monomer unit of the polymeric chain vary from $3.3 \cdot 10^{-4}$ $(l/mol.s)^{1/2}$ in the case of TEG to $2.5 \cdot 10^{-3}$ $(l/mol.s)^{1/2}$ in the case of PEG-14000. The values of k_6 evaluated with the use of the "standard" inhibitor 2,4,6-tritretbutylphenol [12] decrease from $4 \cdot 10^8$ l/mol.s (TEG) to $4 \cdot 10^7$ l/mol.s (PEG-14000).*

Autoxidation of PEG-1500 was studied in o-dichlorobenzene solution at 95° C and under atmospheric pressure of O_2, the PEG concentration was also equal to 10 mol/l. The kinetic curve of oxygen absorption at small degrees of PEG-1500 transformation is of the form $[O_2]_{abs} = at + \varphi t^2$ typical for autoxidation of organic substances. The initial rate of oxidation a equals $(1 \pm 0.2) \cdot 10^{-6}$ mol/l.s, the autoacceleration factor φ is equal to $(8 \pm 1) \cdot 10^{-10}$ mol/l.s^2. The amount of resultant hydroperoxides calculated by the iodometric titration technique virtually coincides with the amount of the absorbed oxygen ($\Delta[ROOH] \approx \Delta[O_2]_{abs.}$, which also correlates with the generally accepted mechanism of chain oxidation of organic substances in the liquid phase.

* The attempt to quantitatively evaluate the role of the "alcohol" peroxyradicals reactions upon PEG oxidation with the use of n-benzoquinone [13] proved to be a failure since in the viscous systems under investigation quinone predominantly reacts with alkyl radicals R rather than "alcohol" peroxyradicals RO_2.

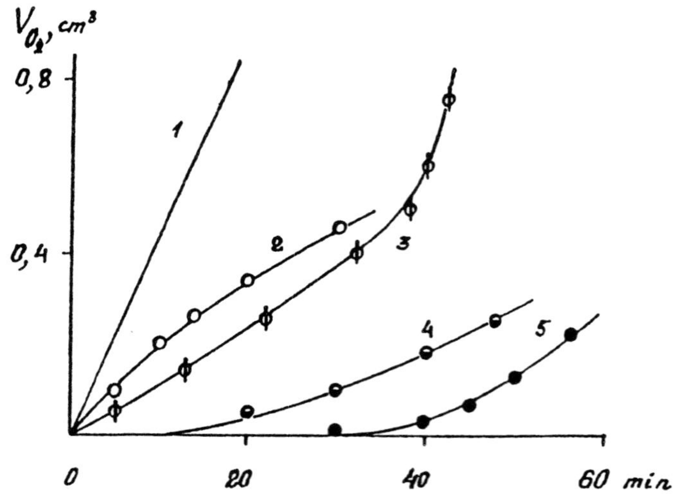

Figure 1. Kinetic curves of oxygen absorption in the initated oxidation of TEG 975°, W_i - 7.5 · 10^{-7} mol/l.s.): (1) without chelate additives, (2–5) in the presence of chelates, (2)—II (1.2 · 10^{-3} mol/l), (3)—XXVI (7.6 · 10^{-4} mol/l), (4)—XXV (1.2 · 10^{-4} mol/l), (5)—XXIV (3.1 · 10^{-4} mol/l).

2.2 Oxidation of Polyethylene Glycols in the Presence of Metal Complexes

2.2.1. Oxidation of Oligomers (TEG and PEG-400)

In initiated oxidation of oligomeric PEG in the presence of all compounds under study the initial rate of oxidation decrease differently, that is, these compounds deactivate the chains leading free radicals. Figure 1 shows typical kinetic curves of oxygen absorption in TEG oxidation. In most cases (particularly in TEG oxidation), in the presence of Cu(2$^+$) and Co(2$^+$) chelates there is no distinct completion of the inhibition period t resulting in reaching the rate of non-inhibited oxidation (with the exception of copper chelate XXII and zinc and bismuth dithiocarbamates XXVI and XXVIII). The most effective inhibitors of TEG oxidation, in addition to XXVII, were found to be copper and cobalt chelates XXIII, XXIV and XXV.

The TEG oxidation is distinctively characterized by considerably long periods of inhibition in the presence of some chelates of copper and cobalt and, accordingly, by higher stoichiometric inhibition coefficients f_t for these complexes. Thus, in the case of XXIV and XXY f_t = 5 to 6, and for IX it reaches 7.5, which is likely to indicate the realization of redox-mechanism of repeated chain termination with the participation of "alcohol" peroxyradicals (reactions 1 and 2) in the presence of the channel of irreversible consumption of metal complexes in the system.

In the case of zinc chelate XXVI (zinc is not prone to single-electron redox reactions) the value $f_t \approx 2$, which agrees with the mechanism of "ligand expelling" proposed earlier for oxidation of hydrocarbons [10]:

$$ZnL_2 + RO_2 \rightarrow ZnLRO_2 + 1/2L_2 \quad (3)$$

$$ZnLRO_2 + RO_2 \rightarrow \text{products} \quad (4)$$

In transition from TEG to PEG-400 the value f_t decreases from 4 to approximately 2 for cobalt chelate VIII and also to 3.5 for copper chelate XXIV. This fact is likely to demonstrate the decreasing role of the reaction of "alcohol" peroxyradicals in the gross process of PEG oxidation upon the increasing degree of polymerization. At the same time, the value of stoichiometric coefficient for zinc chelate XXVI retains its value, f_t = 2.

2.2.2. Oxidation of High-Molecular PEG-14000 and DMEP-5000

In oxidation of PEG-14000 and DMEP-5000 the reactions of "alcohol" peroxyradicals should be either significant or do not proceed at all; in this case the leading chains "ether" peroxyradicals do not exhibit reducing properties.

Figure 2 shows the characteristic kinetic curves of oxygen absorption in oxidation of DMEP-5000 in the presence of chelates (the curves in oxidation of PEG-14000 have a similar shape). Sulfur-containing chelates of zinc and bismuth XXVI and XXVII inhibit the oxidation process. In this case, the maximal values of stoichiometric inhibition coefficient possible within the framework of the simple mechanism of "ligand expelling" are realized: f_t = 2 (XXVI) and 3 (XXVII).

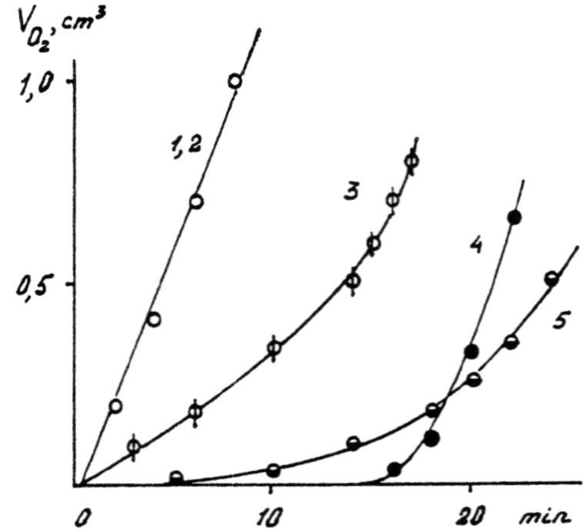

Figure 2. Kinetic curves of oxygen absorption in the initiated oxidation of DMEP-5000 (75° C, w_i - 7.6 · 10^{-7} mol/l.s.): (1) without chelate additives, (2–5) in the presence of chelates, (2)—XVII (6 · 10^{-4} mol/l), (3)—XXVI (4 · 10^{-4} mol/l), (4)—XXY (6 · 10^{-4} mol/l), (5)—XX (6 · 10^{-4} mol/l).

As to the chelates of the variable valence metals, they may be classified into two groups: (1) the compounds completely losing their inhibiting properties in oxidation of PEG-14000 and DMEP-5000 and (2) the compounds retaining their inhibiting activity. The first group includes all tricyclic chelates of cobalt with the active center $Co(N_4)$ (VIII-XII), bicyclic copper XXIV chelate with the $Cu(N,NH)_2$ active center and most of copper bicyclic chelates with the $Cu(O,N)_2$ active center, with the only exception of compounds XIX and XX containing strong electron donor substitutes in ligands. We can assume that the compounds of the first group are not able to terminate the oxidation chains according to the reaction of "ether" peroxyradicals; their inhibiting properties in the process of oligomeric PEG oxidation may be attributed to the interaction with "alcoholic" peroxyradicals.

In addition to the above compounds XIX and XX, the second group of chelates includes bicyclic sulfur- and nitrogen-containing chelate of copper XXII, all bicyclic and tricyclic copper chelates with $Cu(N,N)_2$ and $Cu(N_4)$ active centers and cobalt XXV chelate with the $Co(N,NH)_2$ active center. Alongside with dithiocarbamates of zinc and bismuth, these complexes terminate the oxidation chains, interact with "ether" peroxyradicals; in oxidation of oligomeric PEG these chelates are likely to interact with peroxyradicals of both types. In any case, the nitrogen-containing chelate of copper I actively inhibits the initiated oxidation of cyclohexanol which proceeds with the participation of α-oxycyclohexyl-peroxyl radicals.

2.2.3. Oxidation of PEG-1500

In oxidation of PEG-1500 which, in terms of polymerization degree, is an intermediate between oligomers and high-molecular PEG, compounds of all the classes studied exhibit the inhibiting action of different degree. The characteristic kinetic curves of oxygen absorption in PEG-1500 oxidation inhibited by chelates are given in Figure 3.

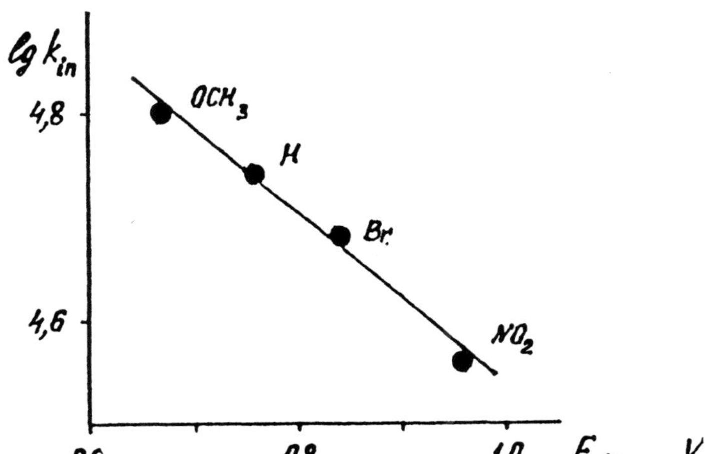

Figure 3. Kinetic curves of oxygen absorption in the initiated oxidation of PEG-1500 (75° C, W_i - 7.4 × 10^{-7} mol/l.s.): (1) without chelate additives, (2–6) in the presence of chelates, (2)—I (6 · 10^{-4} mol/l), (3)—IV (3 · 10^{-4} mol/l), (4)—VIII (1 · 10^{-3} mol/l), (5)—XIII (3 · 10^{-4} mol/l), (6)—IX (1.9 · 10^{-4} mol/l).

Sulfur- and nitrogen-containing chelates of copper XXII, nitrogen-containing bicyclic chelate of copper XIII, bicyclic nitrogen-containing chelate of cobalt XXV and dithiocarbamate of bismuth XXVII appear to be the most active inhibitors, their presence completely suppressing the chain oxidation. The value of stoichiometric inhibition coefficient for compound XIII is equal to 2 to 3 and increases with an increase of the initial concentration of C_0 inhibitor, for chelate XXV the value f_t is equal to approximately 2 and $f_t = 3$ for dithiocarbamate of bismuth XXYII.

Tricyclic nitrogen-containing chelates of copper I, IV and V, bicyclic nitrogen-containing chelate of copper XIV as well as chelates with oxygen- and nitrogen-containing active centers XIX and XX appear to be less active inhibitors.

The values of stoichiometric inhibition coefficients for these chelates are equal to approximately 2 with the exception of tricyclic nitrogen-containing chelate of cobalt IX when the value f_t is significantly higher: at $C_0 = 2 \cdot 10^{-4}$ mol/l $f_t \simeq 5$. The rate of inhibited oxidation in the presence of IX is practically constant during the entire period of inhibition (curve 6 in Figure 3), which agrees with the redox mechanism of chain termination (reactions 1 and 2).

Using the initial rates of PEG-1500 oxidation inhibited by chelates we determined and evaluated the values of inhibition constants fk_7 which quantitatively characterize the ability of these complexes to terminate the chains according to the reaction with free radicals [7]. The initial rates of inhibited oxidation (when it was possible to measure them) were found to be practically unchangeable as the oxygen content in the oxidizing gas mixture changes from 20 to 100%, which indicates the presence of the chain termination according to the reaction of the complexes with peroxyl radicals RO_2. Table 2 gives the values of fk_7 calculated from experimentally found parameter fk_7k_6 with the use of the value $k_6 = 1.10^8$ l/mol.s. For some chelates the values of fk_7 were also determined in oxidation of PEG-400 and DMEP-5000 (the values of k_6 to $3 \cdot 10^8$ l/mol.s and $7 \cdot 10^7$ l/mol.s, respectively, were used in calculations).

In the case of some chelates which do not inhibit the oxidation of DMEP-5000 (that is, they are inactive in reaction with "ether" peroxyradicals) the inhibiting action in PEG-1500 oxidation is due to their interaction with "alcoholic" peroxyradicals. In this case the experimentally obtained inhibition constant is an effective value and it s described by the expression fk_7 exp $\simeq \alpha \cdot fk_7$, where a is the coefficient representing the share of the reaction of "alcoholic" peroxyradicals in the gross-process of oxidation. In the case of all other chelates the experimental value of the inhibition constant seems to represent the total activity of the complexes in reactions with peroxyradicals of both types.

Table 2. Values of inhibition constants fk_7 for chelates in oxidation of PEG-1500 (75° C).

Chelate	$fk_7 \times 10^{-4}$ l/mol.s.	Chelate	$fk_7 \times 10^{-4}$ l/mol.s.
I	1.5	XY	4
II	0.3	XYI	5.3
III	1.1	XYII	6
IY	4.9	XYIII	4.5
Y	6.3	XIX	^{3}10
YI	2.5	XX	^{3}10
YII	2.1	XXI	4
YIII	5.0	XXII	^{3}15
IX	30	XXIII	8
X	13	XXIY	9
XI	8.5	XXY	^{3}15
XII	5		
XIII	^{3}15	XXYI	8
XIY	7.5	XXYII	^{3}15

The data given in Table 2 distinctly show the sharp increase in the inhibiting activity of zinc and bismuth dithiocarbamates XXVI and XXVII as compared to their action in hydrocarbons oxidation: the values of fk_7 in ethylenebenzene at 75° C are equal to $2.4 \cdot 10^3$ l/mol.s and $7 \cdot 10^3$ l/mol.s, respectively [10]. This is the evidence of high activity of PEG peroxyradicals in reactions with chelates as compared with hydrocarbon peroxyradicals. The values of fk_7 for chelate of Zn(2+) (XXVI) practically do not change in transition from PEG-1500 to PEG-400 and DMEP-5000.

The inhibition constants in transition from PEG-1500 to PEG-400 in the case of copper(2+) chelates sharply increase (see Notes to Table 2). This phenomenon seems to result from a considerably more important role of the reactions of "alcoholic" peroxyradicals in oligomeric PEG.

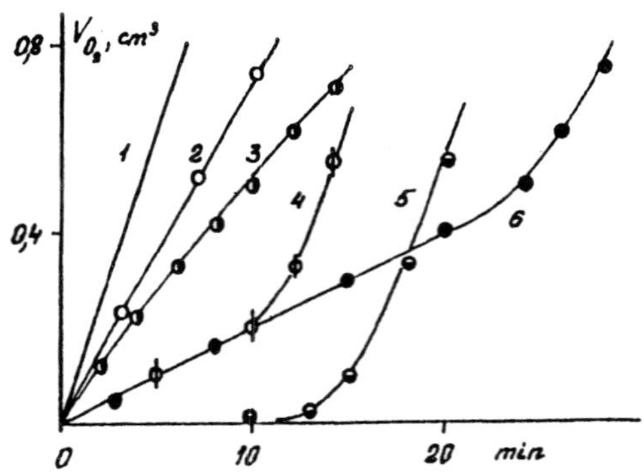

Figure 4. Relationship between the inhibiting activity and electrochemical characteristics of chelates of Co IX-XII (1) and Cu XV-XVIII,XXII (2).

The comparison of the data given in Tables 1 and 2 shows that in some cases there is a correlation between the antiradical activity of metal chelates and their single-electron redox properties (Figure 4). Among similar compounds of cobalt(2+) which is a single-electron reducing agent (chelates IX-XII) one can observe a regular increase in the value of fk_7 as the halfwave potential of oxidation of the complexes ($E_{1/2\ ox.}$) decreases. This fact is an evidence in favor of the inhibition mechanism which includes the redox interaction of the complex with peroxyradicals: Co(2+) + RO_2 → Co(3+) + RO_2^-. As to copper(2+) chelates with bisoxaaza-chelate active centers XV-XVIII, XXI, XXIII, which are inactive in reaction with "ether" peroxyradicals, the values of fk_7 increases as the single-electron reduction of complexes proceeds more readily. This fact completely agrees with the idea of the inhibiting action of these complexes on account of the regeneration by "alcohol" peroxyradicals" Cu(2+) + RO_2 → Cu(1+) + molecular products.

In the case of nitrogen-containing copper(2+) chelates which inhibit the oxidation of both PEG and DMEP-5000 the polar substituents in the ligands change both $E_{1/2}$ and the values of fk_7. Thus, for similar tricyclic complexes IV-VII the electron donor substituent OCH_3 decreases $E_{1/2\ ox.}$ and simultaneously considerably increases fk_7, while the acceptor substituents Br and NO_2 increases $E_{1/2\ ox.}$ and decreases antiradical activity, respectively (Figure 5). The inhibiting action of these compounds seems to be based on the reaction of their oxidation by peroxyradicals. The addition of the strong electron-acceptor substituent CF_3 similarly affects $E_{1/2\ ox.}$

and antiradical activity of tri- and biscyclic nitrogen-containing copper(2+) complexes (compare compounds I and II, IV and III, XIII and XIV).

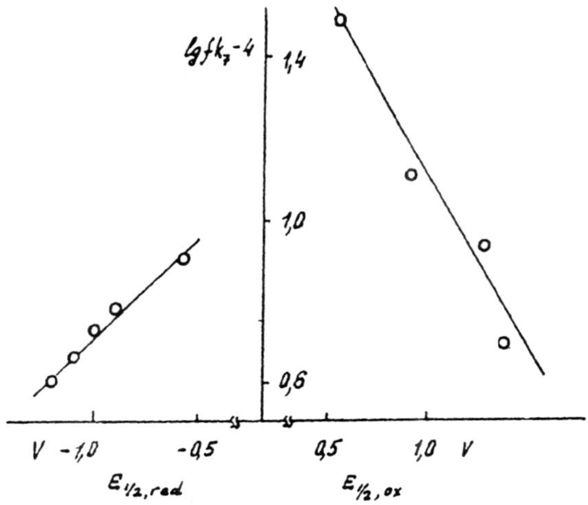

Figure 5. Dependence of inhibition constants on halfwave potentials of electrochemical oxidation of chelates IV-VII (75° C, oxidation of PEG-1500).

However, it should be noted that redox characteristics of the complexes do not entirely determine the inhibiting activity of chelates. For example, in a series of related compounds of copper(2+) with tetraza-active centers I, IV, and XIII varying only in the structure of ligands a considerable growth of inhibiting activity is observed at practically constant value of $E_{1/2\ ox.}$.

2.2.4. Autoxidation of PEG-1500

Figure 6A gives the characteristic kinetic curves of oxygen absorption in autoxidation of PEG-1500 in the presence of chelates. It is seen that the addition of the complexes in the concentrations 10^{-4} to 10^{-3} mol/l has a substantial effect on the oxidation kinetics. All the compounds studied can be divided into three groups:

1) sulfur-containing chelates XXII and XXIV which completely inhibit oxidation;

2) compounds of copper(2+) with $Cu(O,N)_2$ active center, except for chelates XIX and XX, which catalyze oxidation, thus increasing the initial rate and the autoacceleration factor;
3) all the other complexes which catalyze oxidation at low C_0 and exhibit inhibiting properties at which C_0. The characteristic dependences of the initial oxidation rate on C_0 is exemplified by chelates XIII and IV in Figure 6B.

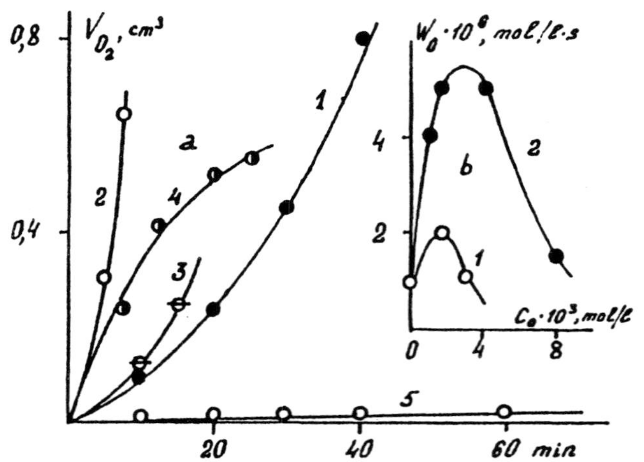

Figure 6. **A.** Kinetic curves of oxygen absorption in the autoxidation of PEG-1500 (95° C, pO_2 to 0.1 MPa): (1) without chelate additives, (2–5) in the presence of chelates, (2)—XXIII ($1.2 \cdot 10^{-4}$ mol/l), (3)—I ($1.2 \cdot 10^{-4}$ mol/l), (4)—XXV ($1.2 \cdot 10^{-3}$ mol/l), (5)—XXII (1.2 $\cdot 10^{-4}$ mol/l).
B. Dependence of the initial rate of the autoxidation of PEG on the chelate initial concentration: (1)—XIII, (2)—IV.

The initial oxidation rate increases in the presence of copper(2+) complex XXIV with the $Cu(N,NH)_2$ active center. However, in the course of the process the rate abruptly drops and becomes less than the initial rate of PEG-1500 autoxidation in the absence of chelates. Preheating of chelate in a solvent in the presence of O_2 decreases the initial rate of catalyzed oxidation and the addition of chelate in the preoxidized solution of PEG-1500 markedly increases this rate. All these phenomena may result from the yield of highly efficient inhibitors, the products of the chelate interaction with O_2 and/or hydroperoxides.

Related nitrogen-containing complexes of copper(2+) I, II and XIII differing in the structure of ligands differently exhibit their catalytic activity. If the addition of I in the concentration of C_0 1.2 · 10^{-4} mol/l does not practically affect the initial oxidation rate, then in the presence of II in the same concentration it abruptly increases and is equal to 4 · 10^{-6} mol/l.s. In the case of bicyclic chelate XIII, the oxidation rate has the intermediate value 2 · 10^{-6} mol/l.s., with the chelate exhibiting, however, the inhibiting activity along with the catalytic one since at C_0 = 3 · 10^{-4} mol/l chelate XIII completely inhibits the oxidation (catalysis of oxidation is observed for I and II at C_0 = 3 · 10^{-4} mol/l).

As a rule, in the case of chelates-catalysts symbasis is observed in the change of the initial oxidation rate *a* and the factor of auto acceleration φ as C_0 complexes change. Thus, it is possible to assume that an increases in *a* results from the rise of initiation rate owing to the decomposition of hydroperoxides present in the initial PEG-1500.

Under similar conditions of DMEP-5000 autoxidation the values of parameters α and φ are close to the values obtained for PEG-1500. The main regularities established for PEG-1500 hold good for the action of chelates.

In conclusion we can state that the copper chelates XIII and XXII with the active centers composition $Cu(N,N)_2$ and $Cu(S,N)_2$ as well as the cobalt chelate XXY with the active center composition $Co(N,NH)_2$ are found to be effective oxidation inhibitors of PEG with various degree of polymerization. Further investigation is required for obtaining a comprehensive picture of the mechanism of the inhibiting action of complexes in PEG oxidation.

3. Reference

1. **G. Furukava,** and **T. Saegusa**. Polimerizatsiya Aldegidov i Okisei (Polymerization of Aldehydes and Oxides) Moscow: Mir Publishers, 1965, 479 pp. (in Russian).
2. **M. Khiraoka**. Kraun-Soedineniya. Svoistva i Primenenie (Crown Compounds. Properties and Application) Moscow: Nauka, 1986, 365 pp. (in Russian).
3. **D.G. Lee,** and **V.S. Chang**. J. Organ. Chem, 1978, Vol. 32, No. 8, p. 1532.

4. **R.S. Goglev,** and **M.B. Neiman**. In: Chemical Properties and Modification of Polymers (Khimicheskie Svoistva i Modifikatsii Polimerov) Moscow: Nauka, 1964, p. 156 (in Russian).
5. **Ph. Grosborne, J. Seree Le Roch,** and **L. Sajus**. Bull. Soc. Chim., France, No. 5, 1968, p. 2020.
6. **N.I. Gordienok,** and **B.G. Freidin**. Kinetika i Kataliz (Kinetics and Catalysis) Vol. 27, No. 4, 1986, p. 988 (in Russian).
7. **N.M. Emanuel, E.T. Denisov,** and **Z.K. Maizus**. Chain Reactions of Hydrocarbons Oxidation in Liquid Phase (Tsepnye Reaktsii Okisleniya Uglevodorodov v Zhidkoi Faze) Moscow: Nauka, 1965, 256 pp. (in Russian).
8. **A.L. Alexandrov, G.I. Solovyev,** and **E.T. Denisov**. Izv. Akad. Nauk SSSR; Ser. khim., No. 7, 1972, p. 1527 (in Russian).
9. **W.G. Lloyd**. J. Polymer Sci., Vol. A1, No. 8, 1963, p. 2551.
10. **V.G. Vinogradova, A.B. Mazaletskii,** and **A.N. Zverev**. Neftekhimiya, Vol. 27, No. 6, 1987, p. 1612 (in Russian).
11. **G.A. Kovtoun, D.L. Lysenko, A.S. Berenbloum,** and **I.I. Moiseev**. Izv. Akad. Nauk SSSR, Ser. khim., No. 2, 1979, p. 293 (in Russian).
12. **G.A. Kovtoun**. In: Chemical Kinetics and Catalysis (Khimicheskaya Kinetika i Kataliz) Moscow: Nauka, 1979, p. 5 (in Russian).

The Laser Power Homogeneous Pyrolysis: Study of the Various Channels of CH_3CClF_2 Monomolecular Conversion

N.Yu. Mordkovich, Yu.N. Zhitnev, E.A. Tveritiniva, and V.V. Timofeyev
Moscow State University
Moscow, Russia

V.G. Barabanov, G.V. Volkov, and P.V. Mitin
State Institute of Applied Chemistry
St. Petersburg, Russia

Kinetic studies on the CH_3CClF_2 destruction induced by the IR-laser pulsed radiation (1.5 Torr) are carried out in the presence of the SF_6 sensitizer. Using the model of physico-chemical processes, the Arrenius parameters in the after pulse period are determined for dehydrochlorination, (Ea = 55 kcal/mol, A = 5×10^{13} s^{-1}), and dehydrofluorination Ea = 65 kcal/mol, A = 1.3×10^{13} s^{-1}) reactions. We have assessed E_a and A for the HF elimination reaction, (E_a = 66.6 kcal/mol, A = 3.3×10^{13} s^{-1}), with the HCl elimination reaction used as the internal reference. The pre-exponential factors of both reactions are assessed by the RRKM method. All the data obtained are in agreement.

Having applied stationary heating of the reaction mixture by continuous wave (CW) CO_2 laser irradiation of a chemically inert sensitizer, Shaub and Bauer [1] found a specific research technique for a high-temperature chemical process in "cold-wall" reaction vessels, (without the heterogeneous steps of the process). The technique was called Homogeneous Laser Pyrolysis (HLP). Substitution of CW-laser radiation by pulsed radiation significantly increases the possibilities of the HLP technique for both the research of chemical process kinetics and controlling the process. Under pulsed pyrolysis, combining almost instant heating of the system (10^{-6} s) and its rapid cooling (10^{-4} s for several-Torr pressures), the secondary processes can be neglected in determining the kinetic constants of the primary steps [2]. Easily dosaged energy of the laser pulse ensures a required temperature in the reaction system and allows for the preset variations of the reactant conversion degrees per pulse in the lowest degree range, (the analysis threshold sensitivity), to the higher degrees, up to the complete conversion of the initial reactant. Kinetic studies under conditions of both CW- and pulsed pyrolysis require detailed information about the temperature and concentration distribution in time ever the reaction volume. These distributions are computed using a model which is determined by a combination of physical and chemical processes found in the system. The pulsed version also requires an adequate model of system cooling in the afterpulse period [2–6].

A number of researchers have employed another technique, the "chemical thermometer." This technique does not require any definite information about precise temperature distribution [5] and the effective temperature of the system can be determined by analyzing the simultaneous courses of two independent monomolecular processes with well-defined Arrhenius parameters. These parameters of one reaction are determined from the known parameters of the other. Though this technique has some significant limitations [7], in HLP conditions it is suitable for investigating the kinetics of complex molecules conversion, these molecules are capable of reacting via several competing channels in the primary monomolecular step.

Investigation of multichannel reaction kinetics of the classical version of upon "smooth" heating of the reactant from the hot vessel walls is hampered in some cases by the superposition of the catalytic processes as well as by the possible "burning-out" of the reactant in the low-energy channel at the moment the system reaches temperatures of noticeable conversion in the high-energy channel. "Instant" reaching of the preset temperature and rapid afterpulse cooling realized in HLP make it possible to maintain the relation of the product yields, in the channel under investigation, closer to the true relation of the respective reaction rate constants. Thus, (e.g., in [8]), the existence of three channels of monomolecular destruction with different activation energies was established.

Table 1

lg A	E_m, kcal/mol	Torr, K	Reactor Type & Material	Ref
14.8	60	858-903	flow	9
12.7	53.8	873-1061	flow, Ni-Cr alloy	10
14.4	60.1	773-973	flow, quartz	11
12.2	53.2	637-758	static, quartz	12
13.4	56	910-958	flow, quartz	13
12.8	54.4	673-773	static	14

This paper presents the results of the experimental study of CH_3CClF_2 multichannel destruction kinetics under pulsed homogeneous IR-laser induced pyrolysis and a comparison of the experimental data with calculated results for these conversions using the RRKM method.

It is known from papers on CH_3CClF_2 thermal destruction that the initial reactant conversion proceeds in two channels [9_13]:

$$CH_3CClF_2 \dashrightarrow CH_2CF_2 + HCl$$
$$CH_3CClF_2 \dashrightarrow CH_2CFCl + HF \qquad (1)$$

It has been argued that dehydrochlorination (1) of this halogen-substitute ethane is a monomolecular process which dominates the higher conversion degrees [12]. There are some discrepancies in the kinetic parameters of reaction (1) which is given in Table 1. Reaction (2) is highly sensitive to the nature and state of the reactor walls. According to the data in [12], the yield of the dehydrofluorination cannot be reproduced. Thus, the degree of this reaction is equal to 0.5 to 7% with passive reactor walls, while in the non-passive quartz reactor, the CH_2CClF content reaches 30% of the total yield. Because of this, the kinetic data for channel (2) were not obtained.

An investigation of the conversion of excited CH_3CClF_2 produced by the chemical activation technique was carried out in [15]. It was found that the experimental results are best described on the assumption that Ea_1 = 58 kcal/mol, Ea_2 = 57 kcal/mol, $A_1 \sim A_2 \sim 10^{13.5}$. On the other hand, the data on the dissociation of the CH_3CClF_2 molecules, multiphoton-excited by the IR-laser radiation, show that reaction (1) is the most competitive channel for freon-142 conversion. It was shown in [16] that CH_2CF_2 and HCl are the only products of IR multiphoton dissociation (IR MED) of

CH_3CClF_2. However, when the laser beam was focused, resulting in an increased vibrational level of reactant molecule excitation, during 1,1,1-chlorodifluoroethane IR MPD fluorescence of both HCl and HF was observed [17].

1. THE EXPERIMENT

The CH_3CClF_2 (reactant) + SF_6 (sensitizer) mixture in the 1:1 ratio under the total pressure of 3 Torr was heated from a TEA-CO_2 laser ($t_{1/2}$ = 100 ns; n = 944 cm^{-1}). The laser beam diameter was equal to that of the reaction cell (1.5 cm). After radiation, the content of the cell was condensed into a pre-evacuated volume to 10^{-3} Torr, then the condensate was vaporized and chromatographically analyzed. A detailed description of the unit is given in [6].

The dependence of the sulfur hexafluoride absorption cross-section in the presence of 1,1,1-difluorochloroethane on the energy density of the laser radiation was determined in a separate series of experiments.

2. RESULTS

CH_2CF_2 is found to be the main product of the CH_3CClF_2 pyrolysis under pulsed heating. The channel (1) yield varied from 10^{-2}% to 40%, with initiating radiation energy equal to 0.5 to 1.2 J. The CH_2CFCl formation was observed at only 0.6 to 1.2 J, its content varying from 2% to 6 × 10^{-2}%. The experimental data are shown in Figure 1 in the coordinates ln (DC / C_0) ¾ $1/T_{vtr}$, where (DC / C_0) is the conversion degree of the initial reactant in the channel, T_{vtr} is the mean reaction mixture temperature calculated in the approximation of adiabatic conversion of the total sensitizer-absorbed energy into heat.

$$\text{Eabs} = \int_{298K}^{T_{V-T,R}} C_{Vi} \cdot n_i \cdot dt, \quad (1)$$

where C_{Vi} is the molar heat capacity of the i-the component and n_i is its quantity.

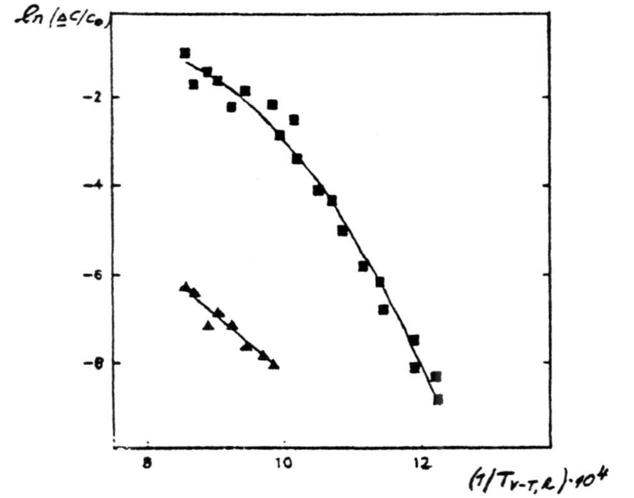

Figure 1. CH_3CClF_2 dissociation yields by channels (1) o and (2) o vs. the initial equilibrium temperature of the system.

Paper [6] presents a rather simple model of the physico-chemical processes which occurring during the course of the afterpulse cooling of the reaction mixture when the laser radiation completely fills up the cell and the heat conduction is the dominating cooling mechanism. The model accounts for the initially inhomogeneous reaction mixture heating due to the non-uniform distribution of energy in the initiating laser beam. Here, the entire cell volume is divided into elementary sub-volumes and the initial energy $E_0(i)$ is assigned to each sub-volume according to their spatial location. Following the well-known relation $dSF_6(E_0)$, *Eabs* (i) and $T_{vtr\,(i)}$ are calculated according to Equation (1). Further temperature variation is determined by computating heat flows between the adjacent sub-volumes. In this case one should take the dependence of heat conduction into account as well as heat capacity on the temperature and the reactant concentration variations in different zones due to both the decay and inhomogeneous temperature field. Thus, a new distribution of temperature and concentration is obtained; this iterative procedure is repeated until the gas mixture cools down to such an extent that the reaction almost stops. The reactant's total conversion degree per pulse ($\Delta C/C_{oclcltd}$) is obtained by summing up the concentration changes over the entire volume and in all time intervals. Thus, to collect the experimental data on ($\Delta C/C_{oexprmtl}$) at different initial energies of the laser beam there exists a system of equations exists

$$\ln(\Delta C/C_0)_{exprmt/(1)} = \ln(\Delta C/C_0)_{c/c/td(1)}$$

$$\ln(\Delta C/C_0)_{eprmt/(2)} = \ln(\Delta C/C_0)_{c/c/td(2)}$$

$$\cdots\cdots\cdots\cdots\cdots\cdots\cdots\cdots$$

$$\ln(\Delta C/C_0)_{exprmt/(n)} = \ln(\Delta C/C_0)_{c/c/td(n)}$$

with two unknown parameters, *Ea* and *A*. There is no precise solution for this system, (due to the natural spread of experimental data), therefore, one should seek its approximated solution corresponding to the FL function minimum for the two variables, *Ea* and *A*,

$$FL = \sum_{k=1}^{N} [\ln(\Delta C/C_0)_{exprmt/(k)} - \ln(\Delta C/C_0)_{c/c/td(k)}]^2$$

and that which yields the best agreement of experimental and calculated data. As shown in [6], the FL function has the only minimum whose coordinates unambiguously correlate with the activation energy and the pre-exponential factor which best agrees with the experimental data on the reactant destruction. The accuracy of these values can be assessed with regard to their dispersion using the standard formulas of regressive analysis [18].

To determine *Ea* and *A* of the dehydrochlorination reaction the destruction degrees of CH_3CClF_2 were calculated. These degrees were related to experiments with low initial temperatures when no dehydrofluorination was observed. Calculations were carried out for the nodes of an alterating-pitch network covering an Ea interval of 40 to 70 kcal/mol and A — 10^{11} to 10^{15} s^{-1}. The minimum was found by using the numerical method with the values, $Ea = 55 \pm 1.5$ kcal/mol and $A = 5 \cdot 10^{13}$ s^{-1}.

The conversion degrees of CH_3CClF_2 in channel (2) were computed in the *Ea* range of 50 to 80 kcal/mol and $A = 10^{11}$ to 10^{15} s^{-1} with respect to reaction (1). The minimum of the FL function for reaction (2) corresponds to the activation energy of 65 ± 2 kcal/mol and the pre-exponential factor of $1.3 \cdot 10^{13}$. Calculation results are shown in Figure 1 as solid lines. As seen in this figure, the shape of the calculated curves completely corresponds to the experimental data. However, as shown in [2,6], a simple graphic determination of the Arrhenius parameters in the aforementioned coordinates is incorrect.

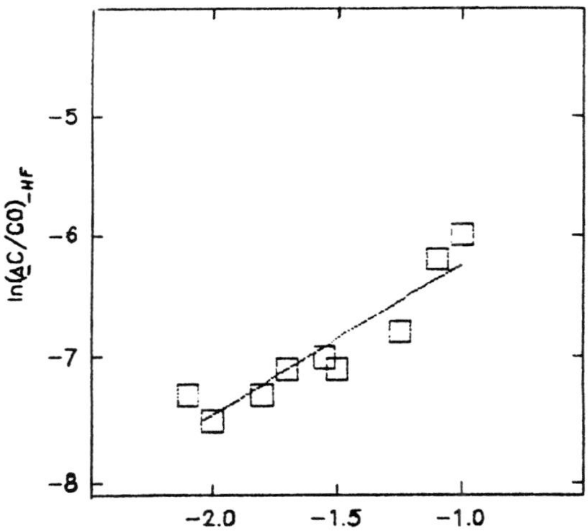

Figure 2. Determination of the Arrhenius parameters of $CH_3CC_1F_2$ dehydrofluorination by the "chemical thermometer" technique. The reference reaction is CH_3CClF_2 dehydrochlorination.

The assessment of *Ea* and *A* of monomolecular process (2) was based on our experimental data and also carried out using the "chemical-thermometer" technique. In this work, channel (1) was used as the internal reference. Here, the following relation should be satisfied:

$$\begin{aligned} LN(\Delta C/C_0)_2 &= LNA_2 - (1-E_1/E_2)\,LN(\tau_r) \\ &- E_2/E_1\,LNA_1 - E_2/E_1\,LN(\Delta C/C_0) \end{aligned} \tag{2}$$

where τ_r is the mean reaction time (it is known from [3,4] that some uncertainty in the assessment of τ_r does not seriously affect the pre-exponential factor determination; in this calculation it was assumed that $\tau_r = 10^{-4}$ s).

The experimental results in the coordinates of Equation (2) are shown in Figure 2. From the slope of the straight line we determined Ea_2 = 66.5 ± 2.5 kcal/mol, and from the section cut at the Y-axis we found A = $3.3 \cdot 10^{13}$.

The values of Arrhenius parameters for reaction (1) are in agreement with results given in the works on CH_3CClF_2 thermal destruction in pyrolysis furnaces [9–14]. The energy barrier for HF elimination from halo substituted ethanes is normally about 60 to 70 kcal/mol [19,20]. Therefore, the value of 65 kcal/mol that we have obtained in our experiments can be regarded as reasonable.

We carried out individual calculations of the rate constants of 1,1,1-difluorochloroethane monomolecular conversion proceeding, according to reactions (1) and (2) on the basis of the RRKM theory [21] using the experimentally obtained results for the energy barrier of these reactions. The assessment of the activated complex (AC) parameters were based on a model suggested by Setser [22] for 4-site elimination of HHal from halo substituted ethanes. The 12 out-of-ring vibrational frequencies were chosen by analogy with the parent molecule and ethane. Four in-plane ring vibrational frequencies were calculated by assigning the bond order of 1.8, 0.2, 0.8, 0.2 to the C—C, C—H, C—Hal, H—Hal, respectively, and then from the Pauling equation [21]. The ring-puckering frequencies were set at 300 cm^{-1}, (reaction (1)), and 500 cm^{-1} (reaction (2)). Calculation yields values of A_1 and A_2 as $10^{13.5}$ and $10^{13.2}$, respectively. These values are smaller than $A = 10^{13.94}$ calculated by the Benson method [23], and yet, quite comparable with the experimentally obtained values.

Calculation of the curves of log $(k_{uni}/k)_\infty$ - log (P) transition for channels (1) and (2) have shown that $P_{1/2}$ is less than 1 Torr even at temperatures above 1000 K. Thus, under the pressure of 3 Torr employed in this research we cannot expect any substantial decrease of the monomolecular reaction rate constant due to a lowered pressure.

The rate constants were also calculated for channel (3) of the chlorine atom detachment from the initial molecule using the RRKM theory

$$CH_3CF_2 - - \rightarrow CH_3CF_2 + Cl \tag{3}$$

The Eo assumed a value of 83 kcal/mol which is approximately equal to the C—Cl bond energy [8]. According to [23], there are two rocking vibration frequencies in AC (equal to 432 cm^{-1} and 309 cm^{-1}) which have isotopic splitting by Cl^{35} and Cl^{37}. They decreased five times and the inertia moment increased four times. Calculation yields $A_3 = 10^{15.1}$. As seen from the ratio of the rate constants for channels (1), (2), (3), channel (3) cannot compete with channels (1) and (2) under these experimental conditions. Calculated results of the kinetic parameters are shown in Table 2, which satisfactorily illustrates the agreement between the results and the literature data (Table 1).

Table 2

Channel	Direct Calculation by Model		"Chemical Thermometer"		RRKM ln A
	E_m, kcal/mol	ln A	E_m, kcal/mol	ln A	
(1)	55 1.5	13.7			13.5
(2)	65 2	13.1	66.5 2.5	13.5	13.2
(3)					15.1

3. References

1. **W.M. Shaub,** and **S.H. Bauer,** *Int. J. Chem. Kinet.,* 7, 509 (1975).
2. **N. Selamoglu,** and **C. Steel.** *J. Phys. Chem.* 87, 113 (1983).
3. **D.F. McMillen, K.E. Lewis, G.P. Smith,** and **D.M. Golden,** *J. Phys. Chem.,* 86, 709 (1982).
4. **H.L. Dai. E. Specht,** and **M.R. Berman,** *J. Phys. Chem.* 77, 4494 (1982).
5. *Laser-Induced Chemical Processes,* (J. Steinfeld, Ed.), Mir Publishers, Moscow, 1984.
6. **Yu.N. Zhitnev, N.Yu. Mordkovich, T. Nugaev, E.A. Tveritinova, V.V. Timofeyev,** and **A.V. Zakharchenko,** *Vestnik MGU* (Moscow State University Papers) ser. *Khimia* 32, 329 (1990).
7. **W. Tsang,** *J. Phys. Chem.* 41, 247 (1964).
8. **D.W. Setzer, T.Sh. Lee,** and **W.C. Danen,** *J. Phys. Chem.* 89, 5799 (1985).
9. **D. Sianesi, G. Neilli,** and **R. Fontanelli,** *Chim. Ind.* 50, 619 (1968).
10. **Yu.A. Panshin,** and **N.M. Panshina,** *Zh.Fiz.Khim.* (*J. Physical Chemistry*) 44, 1396 (1970).
11. **G.J. Marten, M. Godfroid, R. Decelle,** and **J. Varbeyst,** *Int. J. Chem. Kin.* 7, 645 (1972).
12. **G. Huybreshts, Y. Hubin,** *Int. J. Chem. Kin.* 17, 157 (1985).
13. **P.V. Mitin, V.G. Barabanov,** and **G.V. Volkov.** *Kinetika i Kataliz* (*Kinetics & Catalysis*) 29, 1468 (1988).

14. **Z.F. Dong, H. Schneider,** and **J. Wolfrum.** *Int. J. Chem. Kin.* 21, 387–399 (1989).
15. **D.C. Phillips,** and **A.E. Trotman-Dickenson.** *J. Chem. Soc.* (A)9, 1667 (1968).
16. **A.S. Sudbo, P.A. Schultz, Y.R. Shen,** and **Y.L. Lee.** *J. Chem. Phys.* 69, 2312 (1978).
17. **G.A. West, R.E. Weston,** and **G.W. Flynn.** *Chem. Phys.* 35, 275 (1978).
18. *Statistical and Computational Methods in Data Analysis,* (S. Brandt, ed), Mir. Publishers, Moscow, 1975).
19. P. **Cadman, A.W. Kirk,** and **A.E. Trotman-Dickenson.** *J. Chem. Soc. Far. Trans.* 72, 996 (1976).
20. **Yo. Ichikawa, K. Sugita,** and **Sh. Arai.** *J. Phys. Chem.,* 90, 5067 (1986).
21. *Monomolecular Reactions,* P. Robinson and K. Houlbruck, Mir. Publishers, Moscow (1975).
22. **J.S. Hassler,** and **D.W. Setser.** *J. Chem. Phys.* 45, 3246 (1966).
23. **S. Benson.** *The Basics of Thermochemical Kinetics,* Mir Publishers, Moscow (1965).

ORIENTATION EFFECTS ON COLLISION DISSOCIATION WITH ION-PAIR FORMATION

L. Yu. Rusin
Institute of Energy Problems of Chemical Physics,
Academy of Sciences, Moscow
Russia

1. INTRODUCTION

The role of orientation effects in the *CsBr* molecules dissociation induced by collision with *Xe* and *Hg* atoms and proceeding along two channels to yield monatomic ions and molecular complexes $XeCs^+$ is considered. The non-collinear collision with energy transfer through the light atom is the most effective collision for both channels. The interaction of the molecule with the Hg atom results in the atom-ion dissociation upon collision with both atoms. A common dissociation mechanism suitable for describing both channels of product formation in terms of collision kinematics of the initial reagents and the subsequent product scattering has been proposed. Some peculiarities of dynamics of ion_ion recombination in non-equilibrium conditions are considered.

Collision dissociation (CD) of diatomic molecules processes are found to be relatively simple and therefore very suitable for investigating a number of fundamental concepts underlying the implementation of elementary chemical reactions and energy transfer processes. They first of all should include potential energy surfaces (PES) which in the case of CD processes can be diabatic, non-diabatic or adiabatic surfaces depending on the structure of dissociating molecules. Moreover, the CD processes exhibit such peculiarities which make them attractive for investigation. For example, two initial particles yield three free particles through one of the dissociation channels. In this case, the product scattering differs in its character from that in the formation of two particles, which may result in the change of the function excitation product shape near the threshold and despite the strong endoergicity towards the absence of clearly defined threshold behavior. The mechanism of the molecule dissociation includes the process of energy transfer as indispensable stage, the dynamics of this process substantially affecting the dissociation process. In many cases, investigation of the collision dissociation provides detailed information about the energy transfer dynamics. Of special interest is the fact that the CD processes are basically the reactions reverse to recombination. The dynamics of these reactions are not sufficiently studied since it is impossible at present to investigate them with the use of dynamic methods. In this respect the studies on CD contribute to understanding the dynamic peculiarities of the recombination mechanisms. Practically all the available data concerning these mechanisms and factors promoting recombinations are obtained from the experiments "in colb" and consequently do not contain any dynamic information. In this connection, as indicated in [1], the investigation of the CD dynamics and the factors determining the mechanism of these processes would permit more complete characterization of the recombination dynamics in non-equilibrium conditions.

The CD processes of the salts of alkali metals are found to be the simplest and, at the same time, possessing a number of peculiarities. In the general case, these processes proceed along the three channels

$$A + MX \longrightarrow A + M^{-} + X^{-} \tag{1a}$$

$$\longrightarrow AM^{-} + X^{-} \tag{1b}$$

$$\longrightarrow AX^{-} + M^{-} \tag{1c}$$

The family of *MX* molecules contains diatomic particles of various properties in terms of CD [2,3] varying from dissociation only into neutral products to purely ionic dissociation, with the mechanism changing from adiabatic to diabatic. For this reason, these molecules are more attractive for investigation than other molecules, e.g., thallium salts which exhibit the non-adiabatic character of the CD processes. Investigation of diabatic and nonadiabatic processes within one class of compounds makes it possible in rather a simply way to define the topology of diabatic PES and then to generalize it to the nonadiabatic process.

Practical interest in these reactions is explained by the fact that they proceed at energies of values approximately equal to dissociation energy to ~ 10 eV at which the ionization of molecules becomes increasingly important. The range of energies typical for CD corresponds to the conditions realized in MHG generators, high temperature media where CD can be a substantial source of charged or neutral particles.

Identification of the elementary process mechanism is connected with the determination of its dynamic characteristics. The formation of ion-pairs on CD of the salts of alkali metals provides rather rich experimental material including the functions of product excitation, dependences of the reaction cross section on the internal energy of the molecule dissociation, differential cross sections of the product scattering and some other characteristics [4–8] which allow one to quite adequately reduce the potential energy surface (PES) of the process under investigation [6–9].

The latter is of particular importance since in this case it appears possible to investigate in detail the process itself as well as the conditions determining any possible result of collision, e.g., orientation effects influencing the dissociation processes.

Under orientation effects we shall consider the influence of mutual spacing of atoms in the system under investigation at the collision moment and orientation of the velocity vector of the atom-projectile and of the diatomic molecule axis in the mass center system on the collision result.

The role of orientation factors in collinear collisions was considered in Ref. [5] in terms of energy transfer efficiency within the framework of the impulse model. For example, in the system *Xe-Cs-Br* the part of translational energy transformed into internal energy of the molecule is about 60% for the configuration *Xe-Cs-Br*, whereas for the configuration *Xe-Br-Cs* it is about 95%. However further investigations of this reaction [5–11] showed that the role of collinear collisions in the CD process is not very substantial, particularly in the range of energies beyond 7–8 eV, the most feasible orientation depending on various factors.

Further consideration will concern the orientation effects in the CD processes and associated dynamical effects in the process of the formation of active particles. This consideration will be based on the studies devoted to the collision dissociation of *CsBr* molecules with the *Xe* and *Hg* atoms ([7,8,11,13] also see references in these works). These systems are currently studied in more detail. However, the main dependences presented below can be generalized to other *MX*-type molecules, their dissociation into ion pairs being controlled by diabatic PES, e.g., *CsJ, CsCl, CsF* and some other molecules with ionic bond.

The reactions under consideration are characterized by the presence of two channels of collision dissociation: with the formation of atomic ions

$$CsBr + Xe \rightarrow Cs^{+} + Br^{-} + Xe$$

formation of molecular ions

$$CsBr + Xe \rightarrow XeCs^{+} + Br^{-}$$

It should be noted that the cross section of reaction (2a) for *Hg* is bigger than that for *Xe*, while the cross section of reaction (2b) is much higher for *Xe* than for *Hg* [12–14].

Unfortunately, the modern experimental instrumentation is not adequate, with a few exceptions [15], for direct measurement of orientation effects in bimolecular processes despite definite achievements in the development of laser and molecular beam techniques. However, there exists an alternative possibility of at least qualitative study of such fine effects as, for example, the creation of the reliable trajectory model of the elementary process based on adequate potential energy surface. The question about the extent of the model reliability cannot be elucidated in the general form. However, if the criterion of reliability is chosen to be the description with the use of a model of a sufficiently complete set of experimental data and, in particular, topology-sensitive PES of differential (relative to the angle) and double (relative to the scattering angle and the product energy) differential cross sections, then, as shown below, it would be possible to attain a quantitative correlation between the calculated and experimental dependences without any adjustment of the PES parameters obtained in independent investigations. Generally speaking, this approach is not strictly imperative for the processes controlled by PES with complex topology and is quite suitable for PES with simple topology which may include collision dissociation processes.

2. EXPERIMENTAL

The studies on the collision dissociation dynamics were made on the apparatus given in Figure 1, its detailed description being reported in [7,8]. The present section is dealing with a brief description in order to clear up the technique of obtaining the experimental data as well as represent the main experimental dependences. The seeded in H_2 beam of the *Xe* or *Hg* atoms is intersected by the effusive beam of the *CsBr* molecules at the right angle. A differentially heated two-chamber furnace is used as the effusive beam source. Intensity and energy distribution of the supersonic beam are controlled by using the time-of-flight (TOF) spectroscopy and the mass-spectroscopic beam monitor. The effusive beam intensity is controlled by the monitor with surfaceionization. The TOF-mass spectrometer is used as scattering detector. This TOF-mass spectrometer can operate both as a recorder of total cross sections with the selection of products relative to masses and for measuring the

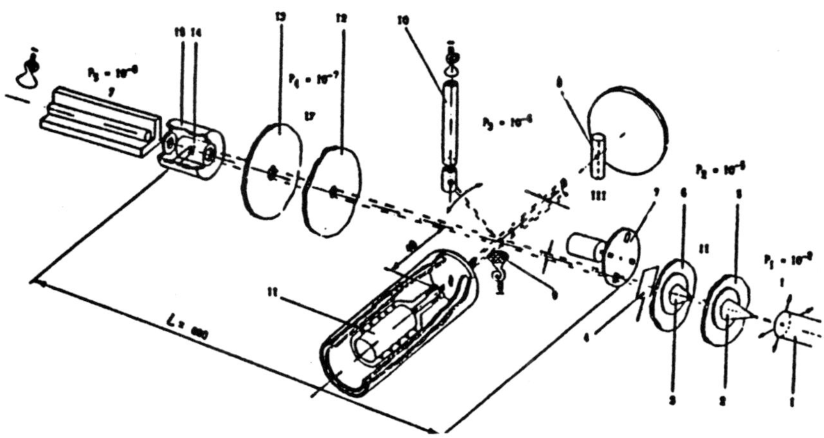

Figure 1. Schematic view of the experimental installation. I–V are differentially evacuated chamber; 1— nozzle; 2—skimmer; 3—collimator; 4—flag indicator; 5,6,12,13—chamber walls; 7—modulator; 8—molecular beam monitor; 9—channel multiplier for measuring the total cross section; 10—scattering detector for time of flight studies; 11—molecular beam source; 14—ionizer for the gas-dynamical beam monitor; 15—cooled shield.

scattered products angle distribution. In this case the mass-spectrometer is shifted from the scattering region at a distance of 50 mm off the center and in front of it is placed an energy analyzer with the retarding field, which enables to obtain double differential cross sections of scattered products.

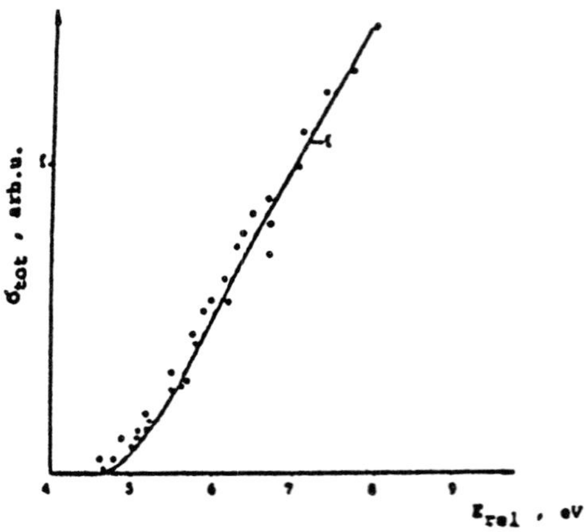

Figure 2. Dependence of the total cross section of the process (2a) on the collision energy. Dots denote the experiment. Solid line is the result of the trajectory calculation after averaging with respect to the velocity distribution in the atom beam.

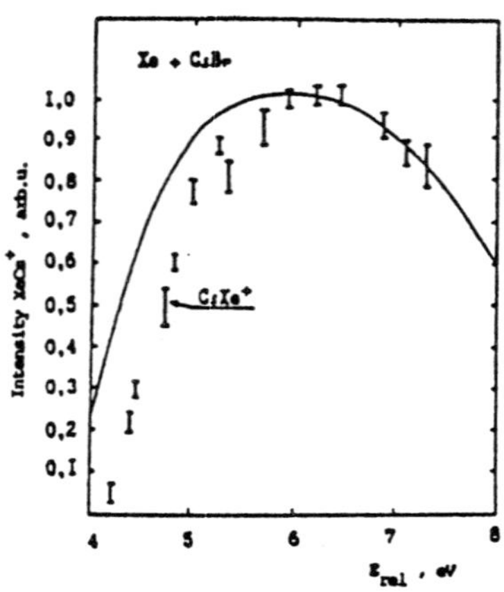

Figure 3. Dependence of the total cross section of the process (2b) on the collision energy. Dots denote the experiment. Solid line shows the result of the trajectory calculation after the averaging relative to the velocity distribution in the atom beam.

The intensity measuring of the scattered product flux and simultaneous velocity analysis and mass separation permit obtaining the data on both reactive channels in identical conditions. The collision energy could vary from the threshold energy (4.6 eV) to 8 eV, i.e., the energy range was within the region which was most interesting for *MX* collision dissociation.

The apparatus described herein is used to obtain numerous experimental data on the dynamics of collision dissociation and, consequently, to test the trajectory model of both the reaction channels. Figure 2 and Figure 3 show the excitation function of channels (2a) and (2b), respectively, for the interaction of *CsBr* with *Xe*. Figure 4 presents the dependence of the total cross section of channel (2a) on the temperature of the molecular beam exit slot, i.e., on the population of the vibrational and rotational states. As seen from the figure, the cross section of process (2b) does not depend on the temperature of the source of the salt molecule beam. The so-called double differential reaction cross sections represented in the angle-velocity plots in the form of contour maps of scattered particles and intensities are most sensitive to the PES topology. The corresponding contour map of the *XeCs* flux scattering is exemplified in Figure 5.

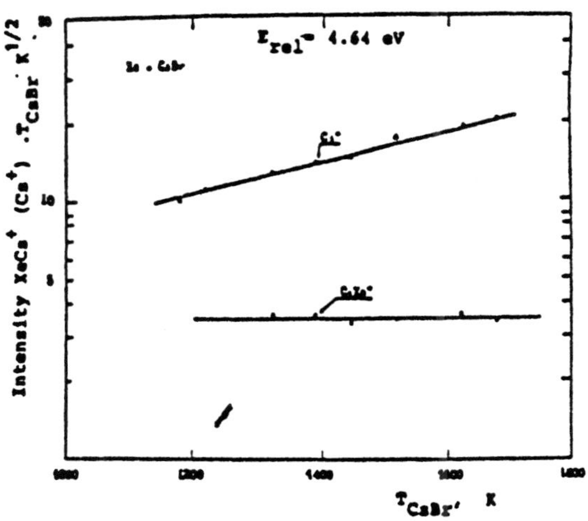

Figure 4. Comparison of the experimentally obtained dependences of the total cross sections of reactions (2a) and (2b) on the temperature of the slot which forms the *CsBr* molecular beam (dots and triangles) with the dependences obtained from the trajectory calculations.

These and other experimental data were used to underlay the trajectory modelling of the CD process along both of the process channels with an aim to identify the PES topology responsible for the process [6–11].

3. Potential Energy Surface

The potential energy surface underlying the model is described in [6–11]. It consists of a sum of pair potentials of the Cs^+ interaction with Br^- and *Xe*(*Hg*) and the interaction with *Xe*(*Hg*). In addition to these members the potential includes non-additive members which describe the induction interactions of the neutral atom with both ions of the *CsBr* molecule. The simple electron structure of the ions and atoms participating in the process enabled the use of truncated Rittner potential for describing the interaction of positive and negative ions and the Mayer-Born approximation of repulsion and interaction of ion-induced dipole $1/R^4$ and dispersed forces *C/R*.

All this results in rather simple PES topology which basically consists in a smooth transition from a relatively narrow valley of reagents to a wide plateau elevated above the valley bed at the depth equal to the value of energy dissociation. The plateau corresponds to the presence of free ions and the atom-projectile. The transfer from covalent PES to ionic PES also proceeds smoothly since the splitting of the relative terms equals ~ 10 eV, and the crossing region is realized over a long internuclear distance RC—27.2 A. The second CD channel is determined by the presence of a shallow valley corresponding to the *AM* ion formation with the energy of bonding equal to 0.1–0.15 eV.

The experimental results are compared with trajectory calculations in Figures 2–6. In Figure 6 the solid lines represent the data from Ref. [4], and the dotted lines designate the trajectory densities. These figures show that in most cases the experimental results completely agree with the calculations despite the fact that the PES parameters were not adjusted and the values obtained from independent calculations were employed (see in survey 10). It should be noted that the trajectory simulation with the recently obtained *Cs-Xe* potential parameters [11] somewhat better describe the experimental data concerning the dynamics of the complex formation than the parameters used in [6], the experimental data on the channels of the complex formation being equally well described. It, for example, refers to the function of complex excitation. This function is sensitive to the potential parameters determining, in particular, the depth of the potential well. It is evident that parameters values used in the simulation are the object of examination both by theorists and experimenters. However, changes in the magnitudes of these parameters do not exceed 10–20%. Thus it is hopeful that further progress in studying the interaction of atoms and ions with closed shells will promote certain improvement of the process model in question.

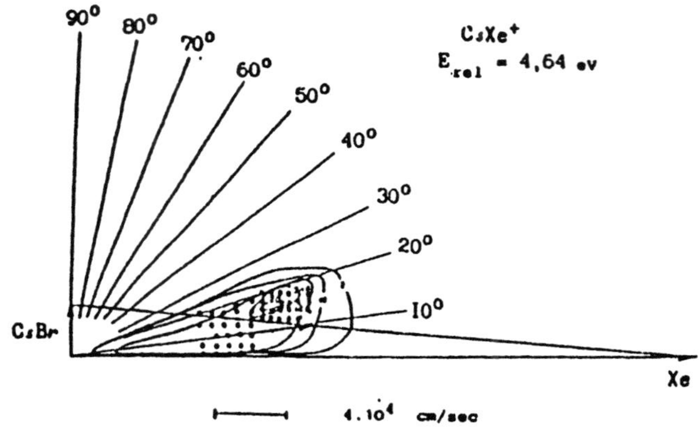

Figure 5. Contour maps of intensity scattering distribution of $CsXe^+$ ions relative to the angle and velocity. Solid lines designate the experiment, dots denote the densities of $CsXe^+$ distributions from the trajectory calculations.

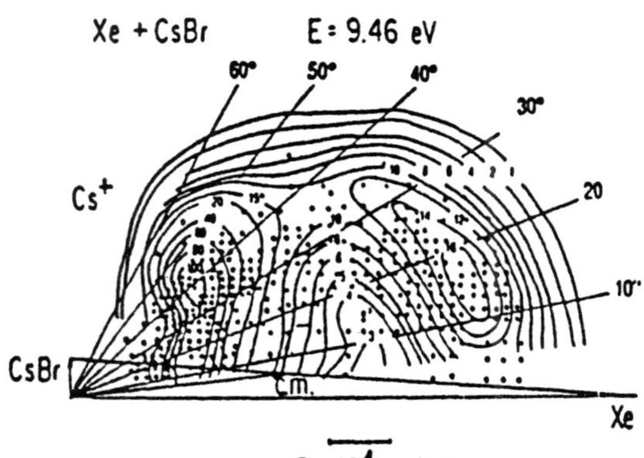

Figure 6. Contour maps of the intensity scattering distribution of Cs^+ ions relative to the angle and velocity. Solid lines denote the experiment [4], dots correspond to the densities of Cs^+ distributions from the trajectory calculations.

The examples given above show that the PES can be used to describe adequately the available experimental data and to interpret the effects unavailable for direct experimental analysis, for example, the role of orientation effects in the CD processes.

4. Main Orientation and Kinematic CD Effects

As said above, the collision dissociation proceeds along two channels which follow the stage of energy transfer. The collision of the atom with the molecule results in the translational energy transfer onto the internal degree of freedom of the molecule and the scattering of the atom-projectile.

The probability of dissociation is mainly determined by the efficiency of the energy transfer onto the internal degrees of freedom E_{in} of the molecule so as to satisfy the condition

$$E_{\text{in}} < D \tag{3}$$

where D denotes the dissociation energy of the molecule along the ionic channel.

The analysis of the simplest models shows that the interaction energy transfer onto the internal degrees of freedom via the interaction of the atom-projectile with the lightest atom of the molecule is, as a rule, the most efficient. This efficiency of energy transfer in the collision of pairs in the framework of the impulse model can be evaluated from the expression

$$E_{\text{trans}} / E_{\text{rel}} - 4m_1 m_2 m_3 (m_1 + m_2 + m_3) / (m_1 + m_2)^2 (m_1 + m_3)^2$$

Here E_{trans} and E_{rel} are the energy transferred to the MX molecule and the energy of collision, respectively; m_1, m_2, m_3 are masses of respective atoms of alkali metal, and halogen.

The expression shows that the highest efficiency of energy transfer for cesium halides in collision with atoms series from *Ar* and *Hg* is attained in the collision of the atom-projectile with the lightest atoms and ranges, in this case, from 0.40 for *Ar-Cs* to 0.98 for *Hg-BrCs*. The same value in collision with the metal atoms varies from 0.122 for *ArCsF* to 0.9 for *Hg-CsJ*. This wide range of changes of the transferred energy part makes the role of the orientation effects understandable.

It is evident that one of the most important orientation effects is the so-called effect of shading the light atom by the heavy one. In this case, prior to collision with the light atom of the molecule the atom-projectile can react with the heavy atom and transfer part of its energy to it. The shading effect substantially limits the region of the angles of mutual orientation of the relative velocity vector and the molecule axis at which the CD process proceeds. It is clear that in the region of very high energies at which the energy transferred via the collision with the heavy atom is sufficient for dissociation the shading effect is not of great importance.

The second type of orientational limitations is also associated with the process dynamics and is of more complex character. These effects will be considered in more detail below.

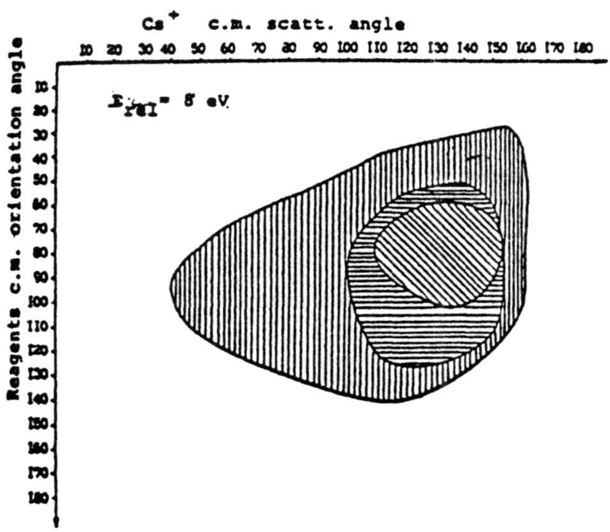

Figure 7. Diagram of relation between the scattering angle of the Cs^+ ions in the center-of-mass system and the orientation angle of the molecule axis and the velocity vector of the *Xe* atoms at the moment of the closest approach. The dash lines show the intensity decrease from the angle towards the periphery.

Different collision geometries may lead to the same result. Figure 7 shows the diagram which connects the angle of mutual orientation of the reagents, i.e., the angle between the atom velocity vector and the molecule axis at the moment of approaching each other, with the angle of the Cs^+ ions scattering in the system of mass center. The *Xe-Br-Cs* configuration corresponds to the angle of incidence 0°, and the *Xe-Cs-Br* configuration corresponds to the angle 180°. The dashed curve from the center towards the periphery shows the decreasing intensities of the product scattering, the difference between the regions being equal to 2_3 arbitrary units. As seen from the diagram, mainly perpendicular collisions or those slightly different from them result in dissociation. As the angle of incidence increases or decreases, the probability for the reaction to approach 0 with the zero target parameter also decreases for collinear configurations *Xe-Br-Cs* and *Xe-Cs-Br*. It is clear that the collinear *Xe-Cs-Br* configuration does not result in a sufficient degree of the translational energy conversion. With the efficiency of energy transfer equal to 60% in this configuration and at the collision energy equal to 8 eV, not more than 4.8 eV, i.e., a little higher than the threshold (4.6 eV), will be transferred to the internal degrees of freedom. The situation with the collinear impact in the *Xe-Cs-Br* configuration is much more complicated. Despite the high efficiency (90%) of energy transfer at the zero target parameter this configuration can lead to a sequence of collisions of the atom-projectile with the light atom and then of the latter with the heavy collision partner in the molecule, which will result in the molecule scattering after the bond tightening. However, this situation occurs at the energies near the threshold or at the energy by 1–2 eV higher. If the collision energy is sufficiently high (e.g., by 3–4 eV higher than the threshold energy), the frontal impact can cause not only a strong tightening of the molecule and its scattering but also its decay into ions.

The non-frontal collinear collision is more efficient. The probability of dissociation increases even at small target parameters of 1–2 at. unit. and the collision energy 6 eV. This probability decreases with a further increase in the impact parameter due to the sliding character of the collision. In this connection, the double collisions of the projectile atom with the target (first with the light atom and then with the heavy one) becomes quite possible. These collisions usually result in the molecule excitation rather than in the molecule dissociation.

Also, collinear collisions can give rise to a situation which might resemble the oscillations of the light atom between the heavy collision partners (see, for example [16,17]). However, in fact, no actual oscillations of the *Br* atom between the heavy atoms occur because of the existing difference in the characteristic oscillation and the collision times despite the fact that the *Xe-Br* distance changes unmonotonously.

As seen from the diagram (Figure 7), the perpendicular or almost perpendicular configuration at the moment of the attack at the *Br* atom or the molecule center is found to be the most feasible configuration for collision dissociation among three mutually oriented reagents, i.e., the collinear and noncollinear orientations during the attack at the *Cs* or *Br* atom. These configuration are most advantageous in terms of the ratio of the participating atoms as well as in terms of the energy transformation relative to the motion of atoms and molecules into the energy of the *Cs-Br* bond breakage resulting in the scattering of these atoms.

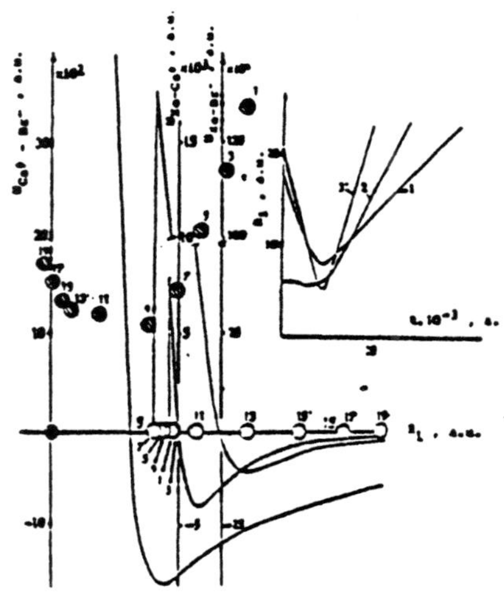

Figure 8. The typical trajectory of the *Xe* atom interaction with the *CsBr* molecule with the impact on the *Br* atom resulting in the molecule dissociation. The onset of calculating the interatomic distances is located in the *Cs* atom center. The curves show the change in the potential energy of the interacting pairs depending on the distances. The insertion indicates the change in the interatom distances during the interaction. All the calculations are performed in atomic units.

In this configuration the energy transfer should first of all result in rotational excitation of the molecule. Dissociation of the latter seems to proceed due to the immediate breaking of the bond during single collision rather than due to the sequence of the T-V-R processes. Figure 8 shows a typical trajectory corresponding to the dissociation in the perpendicular collision configuration. The insertion in the picture shows the dependence of intermolecular distances on time for the same configuration of *Xe-Br-Cs* collisions. It is clearly seen that the process is characterized by the molecule degradation delay (curve 2) as compared to the change in the distance of *Xe-Cs* (curve 1) and *Xe-Br*. In the framework of the impulse model [7] this effect is due to the fact that in the perpendicular collision configuration the collision with the *Br* atom makes the change in the distance between the ions proportional to t^2, i.e., at small values of t the *Cs-Br* distance changes slowly. Another interesting effect is also observed in the perpendicular configuration when the atom attacks the molecule center. In this case, no delay in the molecule decay is observed and the distance between Cs^+ and Br^- starts increasing simultaneously at the moment of the collision partners closest approach. All the intermolecular gaps change simultaneously. In this configuration, expression (4) does not appear to be applicable for the translational energy fraction and the value of the translational energy may be substantially higher.

It should be noted that an increase in the interaction energy widens the range of the angles of mutual orientations resulting in the reaction as well as the range of the angles of the product scattering. The replacement of the projectile atom by a heavier one, i.e., for example, replacing *Xe* by *Hg*, also widens the range of "impact" mutual orientations of the reagents. The diagrams of the product scattering show that in the latter case all the angles from the smallest 20–25° to 130–140° appear to be favorable for reaction (1a), the upper limit being determined mainly by the effect of shading the *Br* atom by the heavy collision partner.

The CD processes involving the *Hg* atom [12–14] exhibit somewhat different orientation effects. In this case, of more importance are collisions of the *Hg* atom with the *Cs* atom, their contribution to the reaction with *Xe* can be neglected in this case. These collisions yield the Cs^+ ions scattering in the forward direction, while the collision with the *Br* atom yields the Cs^+ scattering in the backward direction as in the case considered above. It is obvious that this situation is due to the efficiency of the energy transfer which increases as the projectile atom mass increases.

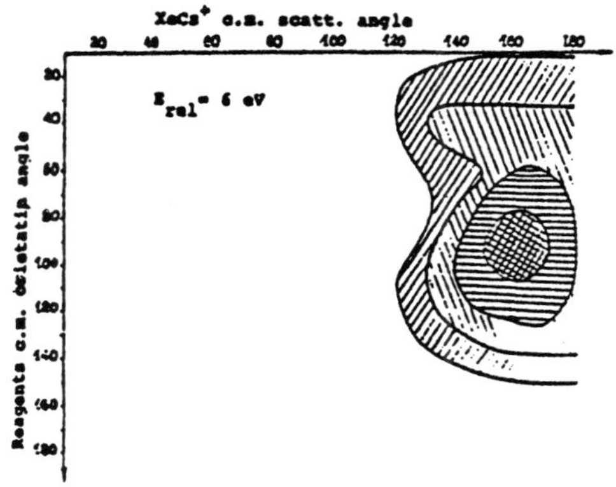

Figure 9. Diagram of relations between the scattering angle of the $XeCs^+$ ions in the center-of-mass system and the orientation angle of the molecule axis and the velocity vector of the *Xe* atom at the moment of the closest approach. The dash lines show the intensity decrease towards the periphery.

The second channel of the collision dissociation (1b) has substantially less cross section but is important for understanding the CD dynamics since the formation of complexes seems to follow the molecule dissociation into ions. Reaction (1b) appears to proceed as the direct exchange reaction and can be classed with the exchange processes of the simplest type. There are two reasons for it. On the one hand, it is due to the predominance of only one type of bonding in the complex, namely, the ion-induced dipole complex, and, on the other hand, due to the fact that the bonding energy in the *XeCs* complex does not exceed 0.15 eV since most part of energy of the system is used for breaking the *Cs-Br* bond and the product translational energy. Only a small part of the remaining energy account for the internal degrees of freedom of the molecular ion due to its weak bonding. As a result, the translational degrees of freedom appear to be the main energy channel. The symmetry found in distribution of the $XeCs^+$ and Br^- products relative to the angle 90° in the system of the masses center and the absence of scattering along the vector of relative velocity appear to be due to the little role which the collinear interactions with a zero or very small target (impact) parameters play in this case. The general picture of orientation effects for this channel can be derived from the diagram of the initial orientation angles, the angle of the product scattering. For the energy 6 eV this diagram is given in Figure 9. As seen from Figure 9, the main part of the product is formed in perpendicular or close to perpendicular orientation of the relative velocity vector towards the molecule axis. Only a small part of the product appears to be formed in collinear or almost collinear collisions with the target (impact) parameters different from zero.

This dynamics of the process probably results from the fact that the complex is formed when the final relative energy E of the Xe atom and the Cs^+ ion does not exceed the strength of complex Q bonding. It means that the value of the relative velocity of Cs^+ and Xe depends on the energy of interaction and the energy transfer efficiency and consequently on the orientation of the interacting particles. The main energetic requirement for the realization of channel (1b) is

$$E_{\text{rel}} < Q, \tag{5}$$

i.e., the complexes will be formed only at such recoil velocities of the Xe atom and Cs^+ ion when their relative velocity compile with this requirement.

In real conditions this requirement is not so strictly observed because the rotation of the complex can considerably distort the interaction potential and the complex can exist at the positive internal energy if the latter does not exceed the magnitude of the centrifugal barrier.

An increase in the collision energy results in two things: the increased backward recoil velocity for Xe after collision and the increased velocity of the Cs^+ ions scattering. This change in relative energy of two particles can be impeding for condition (5). Another obstacle on the way of meeting this condition consists in the fact that the relative energy of Cs^+ and Xe increases not only due to the increasing absolute difference in the magnitudes of these vectors but also due to the increasing angles between them. In other words, a very limited range of collision energies of the initial and the angles of scattering of the resultant ions occurs for meeting condition (5). All the rest set of interaction energies, the reagent orientations yielding in reaction and orientations of the angles between the velocity vectors of the collision products will result in the formation of ions of the atoms. The suggested mechanism of the elementary process of dissociation combines both CD channels into a common mechanism with the same threshold of the process corresponding to the dissociation of the molecule into ions. This mechanism also well accounts for the absence of the formation of the negative complex ion $XeBr^-$ in the reaction of Xe with $CsBr$ since the scattering of Br in the forward direction and that of Xe in the forward direction prevent the fulfillment of condition (5). It should be noted that this mechanism at least accounts qualitatively for those differences in the molecular ion excitation function in other processes with the involvement of cesium halide which were reported in [5].

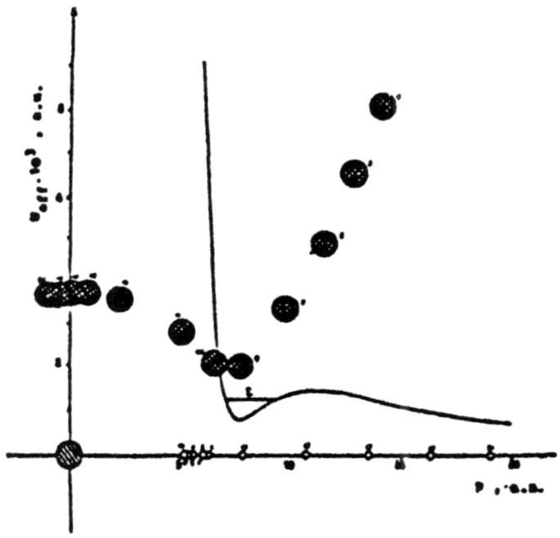

Figure 10. The typical trajectory of the interaction between the *Xe* atom and *CsBr* resulting in the $XeCs^+$ complex formation with the positive internal energy which is less than the centrifugal barrier. The solid line denotes the effective potential of the interaction (in atomic units).

Figure 10 illustrates a typical trajectory leading to the complex formation in collisions close to perpendicular, the total complex energy being close to 0. However, a considerable orbital moment creates the conditions necessary for the stable complex formation. Figure 10 distinctly shows the decreasing velocity of the *Xe* atom motion (positions 9–11) and the resulting small change in the Xe-Cs^+ gap. A similar situation is realized in the interaction of the *Hg* atoms with *CsBr* to yield the $HgCs^+$ complex. However, as said above, the cross section is much less, which is due to the unfavorable ratio of masses of the *Hg* and *Br* atoms since despite a very high degree of energy transfer the relative velocity of *Hg* and Cs^+ at this ratio increases constantly with an increase in the energy interaction.

The presence of both favorable and unfavorable mass ratios of the colliding particles increase the values of the orientational factors not only in terms of energy transfer but also in terms of complying with condition

(5), these factors being decisive in some cases for the realization of channel (1b). The small cross section of the *CsXe*$^+$ formation according to reaction

$$CsF + Xe \rightarrow CsXe^+ + Br^-$$

is most likely to be due to the fact that there action is caused probably by collinear or almost collinear collisions with large impact parameters rather than by perpendicular or almost perpendicular configuration. This interaction may provide a required relative velocity to *Cs*$^+$ and *Xe*, while the noncollinear collisions can result either in the rotational excitation of the molecule or in its degradation into atomic ions. It should also be noted that the *CsF* molecule in the series of cesium halide is an exclusion in terms of collision dissociation because of its structural peculiarities.

One of the interesting aspects of studying the CD dynamics is a possibility of determining some dynamical peculiarities of the ion-recombination processes which are to be the reactions inverse to collision dissociation.

One of the most known ion-recombination models is found to be Thomson model [18] which is somewhat specified and supplemented in [20,21]. This model suggests that formation of a long-lived ion complex of orbiting having an elliptic orbital in which one ion revolves around the other. The appearance of orbiting trajectories is possible within a narrow range of collision energies. A preliminary collision of ions with neutral particles is required to provide this model with the necessary energy of the two particles interaction. Similar collisions stabilize the ion complex. The third particle is to accumulate the excessive energy from the internal degrees of freedom of the resultant molecule. The value of this energy is not specified. The general considerations lead to a conclusion that it should be sufficient enough so as to stabilize the adduct until the next collision occurs. It can result both in the adduct energy offsetting, i.e., further stabilization, or in increasing the adduct energy, i.e., its dissociation. The question of the dynamics of stabilization in triplet collisions remains open in the general case, for many factors responsible for energy transfer remain unknown, such as, for example, the most effective configurations, energetic and dynamical limitations of the stabilization efficiency including the importance of the ratio of the masses of colliding atoms. It is obvious that in terms of this interaction, the recombination process does not appear to an elementary triplet process but includes quite a number of elementary stages finally resulting in the formation of the stable molecule.

Another recombination model is also known [20]. It appeared as a result of experimentally discovered fact of negative temperature dependence of rate constants of a number of recombination reactions. The model suggests the presence of two bimolecular reactions: the formation of the intermediate complex (e.g., the orbitation complex or the labile chemical compound) between one of the recombinating atoms and the third particle and the subsequent exchange reaction between the complex and the second component of recombination. In the simplest case, the ion-induced dipole-potential of this interaction with good approximation can be presented in the following form

$$U(R) = A \exp(-R/\rho) - \alpha/2R^4 - C/R^6$$

It can be assumed that this mechanism will be of certain importance in the field of over threshold energies of ions and at very low pressures or in a highly non-equilibrium conditions. In this case it would be interesting to consider the dynamics of the CD-reverse processes.

It is clear that both CD channels described above correspond to two models or, to be more exact, to two probable mechanisms of recombination (rearrangement)

$$M^+ + X^- + A \longrightarrow A + MX \qquad (7)$$

$$A + M^+ \longrightarrow AM^+ \qquad (8a)$$

$$AM^+ + X^- \longrightarrow A + MX \qquad (8b)$$

In all of these cases, the resultant energy is a same more than the bonding energy in the *MX* molecule (i.e., ~ 4 eV for *CsBr*) but it is not sufficient for exciting the electronic levels of the inert gas atoms. The process of energy redistribution in the system, therefore, is the most important factor for an understanding of the recombination dynamics.

Time symmetry of the classical equation of motion as well as the wave equation in the absence of external fields consists in the fundamental principle of microscopic reversibility [19]. It means that the representative point of the phase space traverses the initial path in the reverse direction when the time sign changes. In other words, the trajectory model constructed for collision dissociation can provide information on the dynamics of reverse elementary processes (of recombination). The properties of CD trajectories give some common characteristics of the dynamical mechanism of recombination in the conditions different from equilibrium ones.

1. Two principal mechanisms of recombination seem to be realized, via, the triplet collision by removal of considerable energy (» kT) and the mechanism of subsequent bimolecular stages (reactions), i.e., via the intermediate diatomic long-lived complex.

2. Both mechanisms of ion recombination are direct and proceed without the formation of long-lived intermediates containing the three particles.

3. Both mechanisms enjoy the most favorable interactions (in terms of choosing the orientation energy) of recombining and stabilizing particles to obtain the final product.

4. Recombination (rearrangement) of heavy ions, for example, the Br^- and Cs^+ ions, with the participation of the third heavy particle or recombination of the complex molecular and monatomic ions proceeds during the times much less than the period of rotation or vibration of the resultant molecule and stabilization of the latter appears to result in the formation of the molecules in the states determined by initial distributions.

The above orientation effects in collision dissociation are mostly due to the effectiveness of the translational energy transfer on the internal degrees of freedom. It is evident that the higher the probability of energy transition to the third member, the higher the molecule stabilization. In accordance with the principle of microscopic reversibility based on the orientation effects in question it is possible to assume that in terms of stabilization the most effective collisions appear to be non-collinear collisions of the third particle with the lightest atom of the resulting molecule or simultaneously with both of them. In these collisions, stabilization does not require the presence of multiple subsequent processes (*R*,*V*-*T*) and, therefore, it may become very important for the media with low gas density. As the gas density increases, the relative role of the most effective configurations will become less important due to the increasingly greater role of collisions with other orientations, for example, such as collinear orientations with a wider variety of impact parameters, collisions with the heavy atom of the resulting molecule, etc.

The triangular configuration seems to be very effective when the energy of both atoms of the molecule is transferred to the atom A translation freedom degrees, i.e., when the stabilization does go through A atom impact with molecule center. The triangular configuration of the collision complex enables also the formation of the M-X bond and then to transfer of some energy in very effective process when dumbbell end hits the third particle transferring of the rotational energy part to the translational freedom degrees of atom A.

Another situation of recombination of monatomic ions consists in the formation of the intermediate or transitional ion pair which is bound by long-range attraction forces, for example, the formation of the orbiting pair which lives for some time until it collides with the third particle. In this case the third particle is likely to affect the state of the pair since the long-range interaction of the Coulomb potential is much greater than in the case of the induction and dispersion forced occurring between the ion and the neutral atom. However, this has nothing in common with the subsequent bimolecular mechanism of recombination described above because the latter includes the formation of the AM^+ or AX^-type long-lived complex followed by the exchange reaction with the X^- or M^+ ions, respectively.

It is clear that the realization of the collinear triplet collisions in the reaction of monatomic ions is of very little probability and the collisions in the non-collinear configuration should prevail in this case.

The realization of the recombination mechanism in subsequent molecular stages and its efficiency are likely to depend on the medium density and the mean particle energy. The association reaction

$$Xe + Cs^{+} \longrightarrow XeCs^{+}$$

followed by the reverse process may become the decisive stage in the range of low pressures. One of the probable mechanisms of the two-body complex formation is found to be orbiting. This recombination mechanism can have a few channels consists in the interaction of the orbiting pair with the ion resulting in the stable molecule due to the presence of the neutral partner of orbitation. The other channel consists in the transformation of the orbiting pair into a stable complex of the AM^+ or AX^- types. Upon the interaction with any third particle. The next stage is reaction of the stabilized complex with the monatomic ion. The rate of the second stage

$$XeCs^{+} + Br^{-} \longrightarrow Xe + CsBr$$

will depend on the dynamical situation as this reaction is affected by very long-range Coulomb forces and it is sufficient to remove a small part of the energy in order to stabilize the molecule. The presence of the heavy third particle in the region of interaction at this moment can make the stabilization process more probable.

Obviously, these two cases do not limit a wide variety of possible recombination mechanisms. It should be noted however that the dynamics of bimolecular recombination as well as the formation of molecular ions in CD can substantially depend on the ratio of the masses of the participating atoms and the energy of the particle collisions.

The peculiar features considered herein for the recombination process are for covering the entire dynamics of recombination of even such simple systems as that of ion recombination. The data on the CD dynamics help elucidate the picture though they do not seem to provide sufficient information on the matter. In the light of these considerations it is reasonable to hope that further progress in studying the CD dynamics of both diabatic and non-diabatic processes and experimental studies of the recombination dynamics will enable development of adequate models of these processes.

5. References

1. **F. Smith**. in *Kinetic Processes in Gases and Plasma* (Kineticheskie Protzessy v Gazakh i Plas my), 1972, 277 p., Russian.
2. **R. Grice,** and **D. R. Herschbach**. *Mol. Phys.,* Vol. 27, 1974, pp. 159–175.
3. **J. J. Ewing, R. Milstein,** and **R. S. Berry**. *J. Chem. Phys.,* Vol. 54, 1971, pp. 1752–1760.
4. **F. P. Tully, N. H. Cheung, H. Haberland,** and **Y. T. Lee**. *J. Chem. Phys.,* Vol. 73, 1980, 4460–4475.
5. **E. K. Parks, G. Pobo,** and **S. Wexler**. *J. Chem. Phys.,* Vol. 80, 1984, 5003–5002.
6. **A. A. Zembekov, A. I. Maergoiz, E. E. Nikitin,** and **L. Yu. Rusin**. *Teoretich. Eksperim,. Khimiya.,* Vol. 17, 1981, pp. 579–586, Russian.
7. **V. M. Azriel, V. M. Akimov,** and **L. Yu. Rusin**. *Khim. Phys.,* Vol. 9, 1900, pp. 1224–1230, Russian.

8. **V. M. Azriel, V. M. Akimov, L. V. Lenin, L. Yu. Rusin.** et al. *Rasseyanie inov* Cs^+ *v protsessakh dissotsiatzii molekul CsBr.* Otchet Instituta energeticheskikh problem khim. fiziki Akad. Nauk SSSR (Scattering of Cs^+ Ions in the Dissociation of *CsBr* Molecules. Report of the Institute of Energy Problems of Chemical Physics, the USSR Academy of Science), Moscow, 1990, No. 0290.0037121, Russian.
9. **V. M. Azriel, V. M. Akimov, J. Gryko,** and **L. Yu. Rusin**. *Traektornoe modelirovanie stolknovitelnoi dissotziatzii solei shchelochnykh metallov v molekulyarnykh puchkakh* (Trajectory Modelling of Collision Dissociation of Alkali Metal Salts in Molecular Beams), VINITI, No. 3574-B89 (Russian).
10. **A. I. Maergoiz, E. E. Nikitin,** and **L. Yu. Rusin** in *Khimiya Plasmy* (Plasma Chemistry) (B. M. Smirnov, ed.), Moscow, Vyp. 12, Energoatomizdat, 1985, pp. 3–55, Russian.
11. **V. M. Azriel, V. M. Akimov, J. Gryko,** and **L. Yu. Rusin**. *Dinamika peredachi energii v stolknoveniyakh Xe s CsBr v diapazone energii ot 0.35 do 11 eV* (Dynamics of Energy Transfer in Collisions of *Xe* with *CsBr* in the Energy Range from 0.35 to 11 eV), VINITI No. 6583-B89, Russian.
12. **L. V. Lenin,** and **L. Yu. Rusin**. *Chem. Phys. Lett.*, Vol. 170, 1990, pp. 502–508.
13. **L. V. Lenin,** and **L. Yu. Rusin**. *Chem. Phys. Lett.*, Vol. 175, 1990, pp. 608–612.
14. **V. M. Akimov, L. V. Lenin,** and **L. Yu. Rusin**. *Chem. Phys. Lett.*, Vol. 180, 1991, pp. 541–544.
15. **G. Scoles**. *Atomic and Molecular Beam Methods,* Vol. 1, Oxford Univ. Press, 1988.
16. **J. Manz, R. Meyer,** and **H. H. R. Schor**. *J. Chem. Phys.*, Vol. 80, 1984, pp. 1562–1568.
17. **E. Pollak, M. Baer,** and **N. Abu-Salbi, D. J. Kouri**. *Chem. Phys.*, Vol. 99, 1985, pp. 15–33.
18. **E. E. Nikitin**. *Teoriya elementarnykh atomno-molekulyarnykh protzessov v gazakh* (Theory of Elementary Atomic and Molecular Processes in Gases), Moscow, Khimiya, 1970, Russian.
19. **V. N. Kondratyev,** and **E. E. Nikitin**. *Kinetika i mekhanizm gazofaznykh reaktziy* (Kinetics and Mechanism of Gas-Phase Reactions), Moscow, Nauka, 1974, 558 p., Russian.
20. **B. H. Mahan,** and **J. C. Person**. *J. Chem. Phys.*, Vol. 40, 1964, pp. 2851–2859.
21. **P. J. Feibelman**. *J. Chem. Phys.*, Vol. 42, 1965, pp. 2462–2470.

A New Autocatalytic Step in the Briggs–Rausher Reaction

V.K. Vanag
Institute of Chemical Physics,
Russian Academy of Sciences
Moscow, Russia

Abstract—The Briggs–Rausher reaction has been studied in the oscillatory and two steady-state regimes — (**I**) low $[I_2]$ and (**II**) high $[I_2$- — under batch conditions at 20 to 27° C in 0.1 M H_2SO_4. It has been found that upon the kinetic phase transition (**I**) → (**II**) I_2 and I^- are autocatalytically produced as the result of decomposition of some iododerivative of malonic acid, R_xI. Under certain conditions the (**I**) → (**II**) transition proceeds not autocatalytically but via a chain mechanism with a constant rate V_{max}. It is shown that R_xI is not iodomalonic acid, and is probably diiodomalonic acid. New mechanisms are proposed for autocatalytic formation of I_2 and I^- that in contrast to the skeleton model of the Briggs–Rausher reaction do not include autocatalytic multiplication of HIO_2.

1. Introduction

The Briggs–Rausher reaction [1], which is oxidation of malonic acid by iodate and hydrogen peroxide catalyzed by the manganous ion, produces bright-yellow to colorless oscillations corresponding to formation and disappearance of iodine, or blue to yellow oscillations in the presence of starch. The skeleton model of this [2,3] and most other oscillatory reactions includes two main sets: a set of reactions ensuring autocatalytic multiplication of some reagent, and a set of reactions providing negative feedback whereby the product of the autocatalytic reactions inhibits autocatalysis.

By analogy with the mechanism of the Belousov–Zhabotinskii reaction, it has been supposed [2,3] that the autocatalytic set is represented by the reactions of autocatalytic multiplication of iodous acid HIO_2:

$$H^+ + IO_3^- + HIO_2 \rightarrow 2IO_2^* + H_2O \tag{1}$$

$$IO_2 + Mn^{2+} + H_2O \rightarrow HIO_2 + MnOH^{2+} \tag{2}$$

The rate of increase in HIO_2 has both a quadratic and a linear limitation:

$$2HIO_2 \rightarrow IO_3^- + HOI + H^+ \tag{3}$$

$$HIO_2 + HOI \rightarrow I^- + IO_3^- + 2H^+ \tag{4}$$

and reaction (5) is also possible:

$$HIO_2 + H_2O_2 \rightarrow HOI + O_2 + H_2O \tag{5}$$

Iodine is formed in reaction (6):

$$HOI + I^- + H^+ \rightarrow I_2 + H_2O \tag{6}$$

The feedback is provided by reactions (4), (7) and (8) producing the inhibitor of reaction (10) — iodide — and reaction (9):

$$I_2 + MA \rightarrow IMA + I^- + H^+ \tag{7}$$

$$H_2O_2 + HOI \rightarrow I^- + O_2 + H^- + H_2O \tag{8}$$

$$H^+ + I^- + HIO_2 \rightarrow 2HOI \tag{9}$$

Along with certain advantages, the proposed mechanism has a number of drawbacks [4]. The phase relations between I_2 and I^- are not adequate to the experiment, Mn^{2+} ions are not oxidized by iodate in 0.1 M H_2SO_4 as they should have been according to the skeleton mechanism, whereas bromate quite readily oxidizes Mn^{2+} ions in conformity with the

Belousov–Zhabotinskii reaction mechanism in reactions analogous to reactions (1) and (2). Edelson [5] has shown that no set of rate constants for the proposed mechanism can provide a correct dependence of the oscillation period on concentrations; moreover, in some cases even the sign of the theoretically predicted change in the period is opposite to that observed in the experiment. These shortcomings are probably caused by disregarding numerous reactions involving organic compounds, in particular interaction of radicals with malonic acid (MA).

To clarify the situation and to elucidate the possibly important but heretofore ignored peculiarities, we decided to study unusual regimes of the Briggs–Rausher reaction whereby such peculiarities could manifest themselves most vividly. It is known [3,6] that in a flow-type reactor the Briggs–Rausher system, like most other oscillatory reactions, can exist in at least four states: oscillatory (**O**), two steady states with low (**I**) or high (**II**) I_2 concentrations, and in a bistable regime (**I**/**II**). On the phase plane in the coordinates $[I_2]_0$, $[IO_3^-]_0$ (the subscript "zero" denoting initial concentrations) these four states are delimited by a cross-shaped diagram [6]. Certain correspondence to these states can also be found for a batch reactor which we use to exclude additional dependences.

To actualize unusual regimes we, for one thing, tried to carry out the reaction on the border of stability of the above mentioned states, and for another, examined the reaction for long enough times in order to observe the possible transitions between these states closer to the end of the reaction and to reveal the possible role played in these transitions by the accumulating reaction products, mainly derivatives of malonic acid.

2. Materials and Methods

Malonic acid (MA), KIO_3, KI, $Mn(NO_3)_2 \cdot 6H_2O$, H_2SO_4, H_2O_2 were of analytical grade; hydrogen peroxide was stabilizer-free.

The course of the reaction was monitored by simultaneously recording $[I_2]$ buy absorbance at 463 nm, $[I^-]$ with a iodide-selective electrode, $[I_3^-]$ and malonic acid derivatives by absorbance at 360 nm, as well as the volume of evolved gas and the potential of a platinum electrode U_{Pt}. The potentials of the iodide-selective electrode U_I and of the platinum electrode were recorded against an Ag—AgCl reference electrode connected with the reaction vessel via a Na_2SO_4-saturated bridge. All the values of electrochemical potentials are henceforth given relative to the standard hydrogen electrode. We used the iodide-selective electrode manufactured by Ecotest, Moscow on the basis of a pressed AgI/Ag_2S tablet. It was calibrated with iodide solutions in 0.1 M H_2SO_4. In the range of iodide concentrations from 10^{-3} to $2 \cdot 10^{-5}$ M the electrode potential U_I logarithmically depend on $[I^-]$. At $[I^{<\ 10}$-5 M, which corresponded to U_I > 437 mV. the electrode sensitivity was substantially reduced and in pure 0.1 M H_2SO_4 U_I = 457 mV, whereas in a reaction mixture 0.0225 M IO_3^-/0.06 M MA/0.61 M H_2OH_2/O.1 M H_2SO_4 it was even higher (ca. 500 mV). Under these conditions the electrode potential is determined by silver ions generated by the corrosion process on the surface of the silver halide membrane [7].

The volume of evolved gas was measured conductometrically with a unit described earlier [8]. Its operating principle is to record the rectified AC passing through a weak salt solution that is displaced by gas and thereby rises in a cylinder where two Nichrome wires are drawn along its sides and an alternating voltage (ca. 1000 Hz) is applied to them.

The changes in iodine and triiodide concentrations were followed using a spectrophotometric unit of our own design [8] which allows the changes in absorbance from 0 to 5 A to be measured with an accuracy of 0.001 A. The unit was modified so that simultaneously two perpendicular beams — monochromatic and white light — could pass through the cell. The monochromatic beam after the cell went again to a monochromator and then to a photomultiplier (FEU-106). The white light beam after the cell went to a monochromator and then to a photodiode array of 1024 elements (Interunis). All analog signals registered were supplied through a 12-bits analog digital converter to an IBM PC AT. Using the photodiode array we could record optical spectra in the 250–750 nm range with a frequency of 58.8 Hz as well as the process kinetics at fixed wavelengths.

Absorptitivities (M^{-1} cm^{-1}) used were: $[I_2]$, e_{463} = 746; $[I_3^-]$, e_{360} = 25,000 and e_{463} = 975; [IMA], e_{350})= 14.8 [9,10] and ϵ_{360}^{IMA}, iodomalonic acid was prepared in solution from a iodide/0.1 M MA/ca. 1 M H_2O_2 mixture at low iodide concentrations (< 0.01 M).

The experiments were conducted in thermostatted batch Teflon cylindrical cells with flat silica windows pressed in on four sides. Electrodes were inserted into the cell lid. The cells had volumes of 10 or 20 ml and optical paths of 2 or 3 cm, respectively. Stirring was mostly done magnetically at fixed rates, using Teflon- or glass-coated stirring bars. In some experiments a Teflon mechanical stirrer was used. The temperature was controlled with an accuracy of ±0.5° C and varied from 20 to 27° C.

Stock solutions were stored in glass-stoppered bottles protected from light.

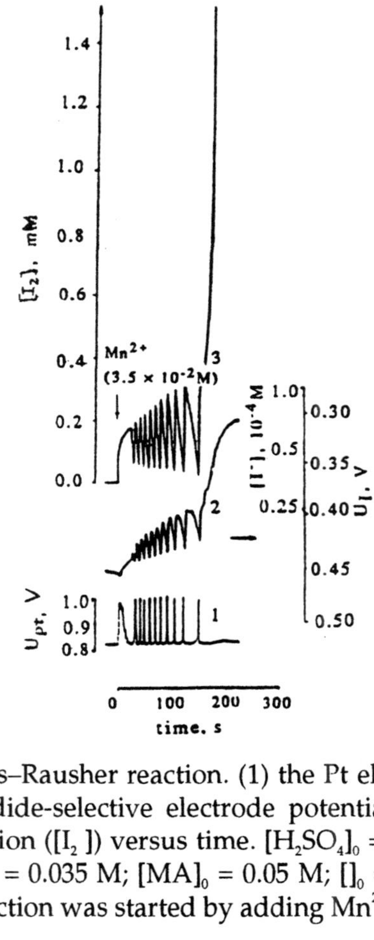

Figure 1. The Briggs–Rausher reaction. (1) the Pt electrode potential, U_{Pt}; (2) the iodide-selective electrode potential, U_{I}; and (3) iodine concentration ($[I_2]$) versus time. $[H_2SO_4]_0 = 0.1$ M; $[H_2O_2]_0 = 0.61$ M; $[Mn^{2+}]_0 = 0.035$ M; $[MA]_0 = 0.05$ M; $[\,]_0 = 0.0025$ M; temp. 27° C. The reaction was started by adding Mn^{2+}.

3. Results and Discussion

To detect unusual regimes and transitions between different states of the system, the reaction is started in the oscillatory mode at the following initial concentrations of the reagents: $[H_2SO_4]_0 = 0.1$ M, $[KIO_3]_0 =$ 0.01 to 0.04 M, $[MA]_0 = 0.02$ to 0.08 M, $[H_2O_2]_0 = 0.5$ to 2.0 M, $[Mn^{2+}]_0 =$ 0.005 to 0.05 M. Depending on the initial concentrations, the oscillatory mode culminates in a transition either to state (**II**) (Figure 1) or to state (**I**) (Figure 2). The most sensitive parameter on which the transition path depends is the ratio of the initial concentrations of malonic acid (MA) and iodate, $a = [MA]_0/[IO_3^-]_0$. At other reagent concentrations chosen, $[H_2O_2]_0 = 0.61$ M and $[Mn^{2+}]_0 = 3.5 \cdot 10^{-2}$ M, the critical value of this parameter a_c is equal to 2.0 at 22° C and 2.2 at 27° C. Exceeding the a_c by just 2 to 3% results in the transition (**O**) → (**I**) whereas at a/a_c the transition is (**O**) → (**II**). State (**I**) of the system is rather sensitive to external influences [11]. The concentrations used henceforth to actualize the (**O**) → (**I**) transition are $[MA]_0 = 6 \cdot 10^{-2}$ M, $[IO_{3_0}^-] = 2.25 \cdot 10^{-2}$ M.

In Figure 2 the arrow Tr marks the end of oscillations and the onset of state (**I**). Let us consider the evolution of state (**I**) in more detail. With vigorous enough stirring (Reynolds number ca. 10,000) the only observed change after reaching a maximum at 360 nm is the slow decline in absorbance (curve 4 in Figure 3), a slow increase in U_I and decrease in U_{Pt}. In the process, the concentrations of I_2 and I^- do not exceed $4 \cdot 10^{-6}$ M and 10^{-6} M, respectively (a more accurate determination is difficult because of bubbles in the cell).

With weaker stirring (Re 10^3 to 10^4), state (**I**) may after some time depending on the stirring intensity spontaneously transform into state (**II**) with an abrupt discharge of I_2, I^-, and (curve 1in Figure 3). The causes of the quite strong dependence of the lifespan of state (I) on stirring intensity will be considered in the next work. We have earlier reported [11] that the (**I**) → (**II**) transition can be initiated by illumination. Reasoning within the framework of the skeleton model, we supposed that the observed transition can happen only owing to the single autocatalytic set of reactions (1), (2) multiplying HIO_2 wherein the source of iodous acid and iodine is iodate. However, a more detailed analysis and additional experiments demonstrate that this is not so.

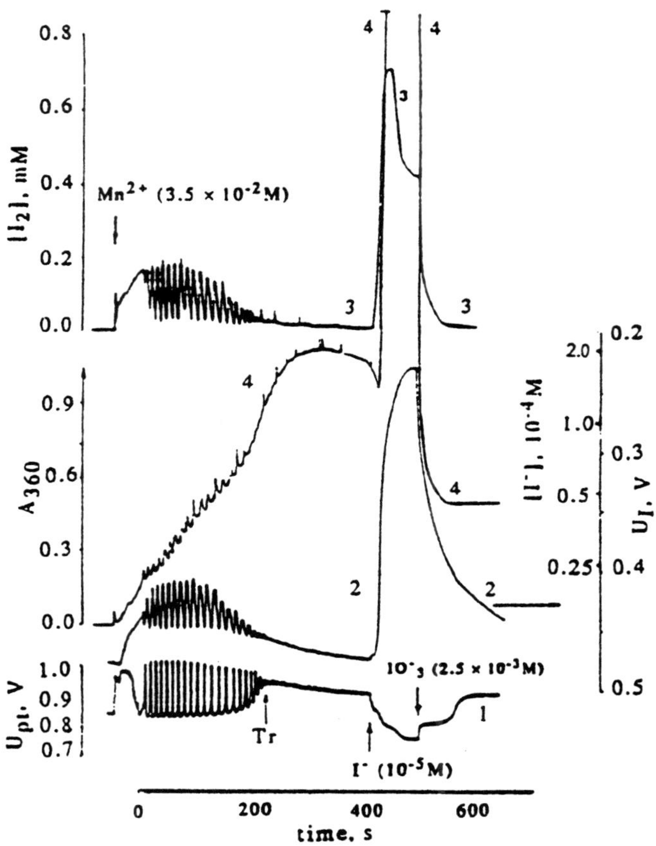

Figure 2. The Briggs–Rausher reaction. (1) the Pt electrode potential, U_{Pt}; (2) the iodide-selective electrode potential, U_{I}; (3) $[I_2]$; and (4) absorbance at 360 nm versus time. $[H_2SO_4]_0 = 0.1$ M; $[H_2O_2]_0 = 0.61$ M; $[Mn^{2+}]_0 = 0.035$ M; $[MA]_0 = 0.06$ M; $[\,]_0 = 0.00225$ M; temp. 27° C. The reaction was started by adding Mn^{2+}. The arrow Tr denotes the transition from the oscillatory to the steady-state (**I**) regime of the Briggs–Rausher reaction. Arrows also mark the moment of addition of iodide that causes a (**I**) → (**II**) transition, and the moment of addition of iodate that causes a (**II**) → (**I**) transition.

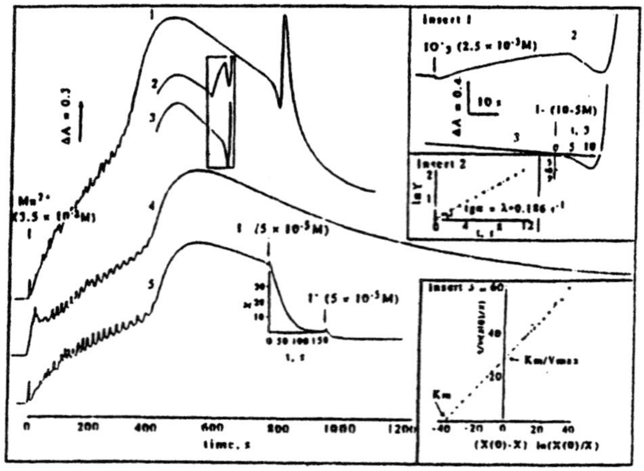

Figure 3. The Briggs–Rausher reaction. Absorbance at (1–3) 350 nm and (4,5) 360 nm versus time. Initial reagent concentrations are the same as in Figure 2; temp. 20° C. Insert 1 represents the framed fragments of curves 2 and 3 with an expanded time scale. Insert 2 shows the dependence of lnY on time, where Y are arbitrary units denoted in Insert 1 and $t = 0$ corresponds to the moment of iodide addition. Insert 3 gives a $t/\ln(X_0/X)$ *versus* $(X_0 - X)/\ln(X_0/X)$ plot of a linear anamorphosis of the integral Michaelis_Menten equation for the rate of a catalytic reaction $dX/dt = -(V_{max}X)/(k_m + X)$ where X are arbitrary units, $X = X_0$ at $t = 0$, and $t = 0$ corresponds to the moment of iodide addition (curve 5).

Table 1. The l values for the exponential growth of $[I_2]$ and the induction periods τ during the kinetic phase transition (**I**) → (**II**) in the Briggs–Rausher reaction presented in Figure 4, as dependent on the concentration of added iodide.

$[I^-]_{add}$, 10^{-5} M	λ, sec^{-1}	τ, sec
0.1	0.186	98.4
0.55	0.246	39.5
1.65	0.191	19.4
3.3	0.224	15.2
5.0	0.229	13.2

It turns out that, first, the (**I**) → (**II**) transition can be initiated by adding iodide which is the inhibitor of steps (1) and (2) (see Figures 2 and 3), and second, by the moment of the transition there is practically no iodate left in the system. In these experiments, iodide is added after the system has passed through the maximum of absorbance at 360 nm, 600 sec from the onset of the reaction (curve 3 in Figure 3). According to the skeleton model [2,3] iodide must suppress the autocatalytic step at $[I^-] > [I^-]_c$ where $[I^-]_c \approx (10^{-5}$ to $10^{-6}) \cdot [IO_3^-]$. In our case the concentration of added iodide is several orders of magnitude higher than the critical one, and nevertheless autocatalysis takes place. If the (**I**) → (**II**) transition is monitored by I_2 formation, there is a clear induction period t, which here is determined as the time from iodide addition to the moment when the transition rate reaches a maximum. It turns out that the higher is the $[I^-]_{add}$ the earlier is the (**I**) → (**I**) transition (Figure 4). With increasing $[I^-]_{add}$ the induction period t decreases according tot he equation $t = const./([I^-]_{add})^{1/2}$ (Table 1). The spontaneous transition under such conditions takes place 870 to 900 sec after the onset of the reaction. Addition of iodide can be said to be analogous to the effect of light on state (**I**) [11]. Just the same conclusion about bromide has been reached by Vavilin et al. [12] with regard to the oscillatory mode of the Belousov–Zhabotinskii reaction. These authors suppose that illumination causes photodecomposition of bromomalonic acid and bromacetic acid with generation of Br^-.

The increases in $[I_2]$ and $[I^-]$ take place concurrently and are described by dependences of the type exp(l*t*), with the exponent l being independent of $[I^-]_{add}$ (Table 1). In order to answer the question whether autocatalytic formation of I_2 and I^- is the result of one process or two linked processes having different rates, we have analyzed the kinetics of the (**I**) → (**II**) transition and determined the l values for I_2, I^-, and formation rates (Figure 5). As shown in Figure 5, iodine first increases exponentially with l = 0.283 sec^{-1}, and when the exponential region changes to linear the value for iodide still remains equal to 0.282 sec^{-1}. The same value is also obtained from the straight portion of the decline in U_{Pt} by differentiating the Nernst equation on assumption that the changes in the Pt electrode potential correlate with $\ln([I_2]/[I^-]^2)$ [4]. For triiodide in accordance with Equation (10):

$$\left[I_3^-\right] = K[I_2][I^-] \quad (10)$$

where K = 704 M^{-1} [9,12], the exponent is 0.54 sec^{-1} ≈ 21 (Figure 5). The exponent depends not only on the reagent concentrations and temperature but also on the stirring rate. This explains some differences between the l values given in Table 1 and the value obtained in the experiment described above.

Gas-volumetric measurements show that gas production stops completely just as the absorbance at 360 nm reaches a maximum; the volume of evolved gas (oxygen) corresponds to that calculated from the stoichiometric equation for the overall process:

$$H^+ + IO_3^- + 2H_2O_2 + MA \rightarrow IMA + 2O_2 + 3H_2) \quad (11)$$

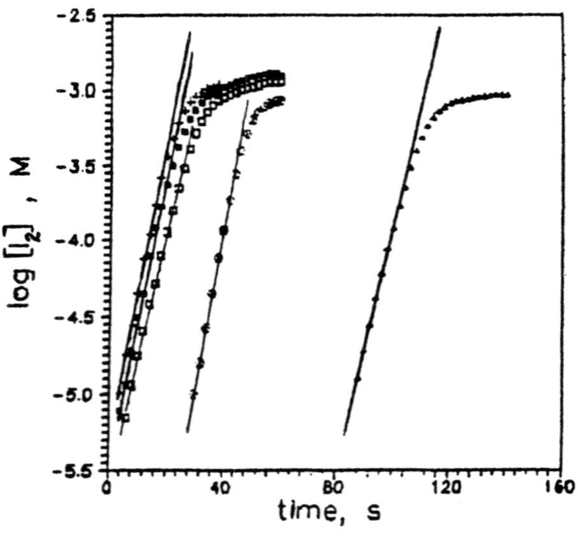

Figure 4. Time dependences of $\log_{10}[I_2]$ characterizing the kinetic phase transition (I) → (II) in the Briggs–Rausher reaction. Initial reagent concentrations are the same as in Figure 2; temp. 20° C; t = 0 corresponds to the moment of iodide addition, and $[I^-]_{add}$ (10^{-6} M) are (▲) 1; (•) 5.5; (❐) 16.5; (■) 33; and (+) 50. Straight lines through the initial parts of the curves were obtained by the method of least squares.

From Equation (11) it follows that under excess H_2O_2 and MA, 2 moles of O_2 should finally evolve per mole . In our case, calculation yields 20.2 ml of oxygen per 20 ml of the reaction mixture, while the measured volume of evolved gas is 21.2 ml. Assuming that the average CO_2 content in the evolved gas is 5% [14], we can say that by the moment of iodide addition there is no iodate left in the system, or at most its concentration does not exceed 10^{-4} M (with allowance for the volumetric error). Similarly, addition of small amounts of iodate instead of iodide to the system 500 to 600 sec after the onset of the reaction invariably results in a rise in absorbance at 360 nm, which in some cases is followed by a spontaneous (**I**) → (**II**) transition (curve 2 in Figure 3); this also testifies to the absence of iodate. Since the concentration of liberated iodine constitutes (1 to 1.5) × 10^{-3} M even disregarding the uptake of iodine by malonic acid during the (**I**) → (**II**) transition, and is several-fold higher if such consumption is taken into account, then the question arising is the source of I_2 and I^- produced in the system.

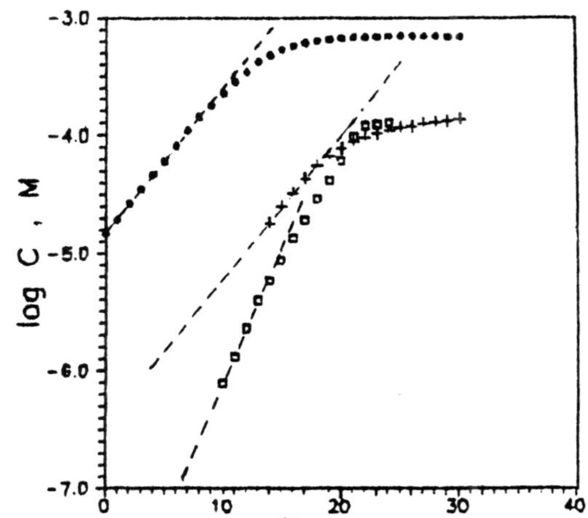

Figure 5. Time dependences of (•) $\log_{10}[I_2]$, (+) $\log_{10}[I^-]$, and (❑) $\log_{10}[I_3^-]$ characterizing the kinetic phase transition (**I**) → (**II**) in the Briggs–Rausher reaction. Initial reagent concentrations are the same as in Figure 2; temp. 25° C; t = 0 corresponds to the moment of iodide addition, $[I^-]_{add} = 5 \cdot 10^{-5}$ M. Straight lines through the initial parts of the curves were obtained by the method of least squares.

It should be noted that the (I) → (II) transition itself is accompanied by liberation of approximately 0.6 ml of gas from 20 ml of the reaction mixture, which corresponds to $1.3 \cdot 10^{-3}$ M of dissolved gas. This value coincides with the concentration of I_2 formed.

After the (I) → (II) transition the system also does not produce iodate, which in principle can be formed in a complex overall reaction of iodine oxidation by hydrogen peroxide and in elementary reactions (3) and (4). Indeed, addition of iodate at a concentration $[IO_3^-]_{add} > [I_2]$ (see Figure 2) results in a rather rapid (characteristic time $(k_7[MA])^{-1}$ sec where $k_7 \approx 40$ M^{-1} sec^{-1} [2–4]) and practically complete disappearance of I_2, I^-, and , which can be attributed to a combination of reaction (7) and the Dushman reaction

$$6H^+ 5I^- + IO_3^- \rightarrow 3I_2 + 3H_2O \quad (12)$$

If after the (I) → transition the system is left alone, I_2 ought to slowly disappear in reactions (6), (7) and (13):

$$H^+ + I^- + H_2O_2 \rightarrow HOI + H_2O \quad (13)$$

with a characteristic time $(2k_{13}[H_2O_2])^{-1}$ sec where $k_{13} \approx 0.04$ M^{-1} sec^{-1} (this work). In reality I_2 disappears still more slowly, suggesting the existence of an additional source of iodine.

Thus, we are forced to admit that there is another autocatalytic reaction distinct from the (1) to (2) set, which produces I_2 and I^- from a source other than iodate. By the moment of initiating the → (II) transition the only source of iodine in our system can be the iododerivatives of malonic acid.

As can be seen in Figures 2 and 3, the phase transition (I) → (II) in all cases begins with a small drop in absorbance at 360 nm, which is autocatalytic (insert 1 in Figure 3) with an exponent $l = 0.186$ sec^{-1} (insert 2 in Figure 3) being close to the constants in Table 1. We believe that this drop just corresponds to decomposition of a iodoorganic compound R_xI that is the source of I_2. The coincidence of the exponents of the process of I_2 and I^- formation with that of R_xI decomposition allows us to conclude that this is one and the same process.

Figure 3, curve 5 depicts a single irreproducible case (which is probably due to uncontrolled deviations from the standard reaction conditions) when after iodide addition there was not autocatalytic formation of I_2, , and I^- but only a decrease in A_{360}. Assuming that the decline at 360 nm corresponds to the decomposition of R_xI, the rate of this process is well described by the general equation for catalytic reactions $V = V_{max}[R_xI]/(k_m + [R_xI])$ (insert 3 in Figure 3). This may mean that either Mn^{2+} take part in this process as a catalyst, or R_xI decomposes via an unbranching chain mechanism that is distinguished by a constant rate V_{max} over a considerable length.

Naturally, we have tried to obtain the compound R_xI and to record the process and analogous to the one depicted by curve 5 in Figure 3 in some other reproducible way. We have managed to do this by-passing the oscillatory mode only in reactions that obligatorily involve iodate. To this end, small portions of iodate (0.0025 M) were added to the mixture 0.06 M MA/0.61 M H_2O_2/ O.00225 M Mn^{2+}/0.1 M H_2SO_4. The system thus obtained (state **I**′) behaves in practically the same way as state (**I**). The increase in absorbance at 360 nm to a maximum is followed by a slow decline.

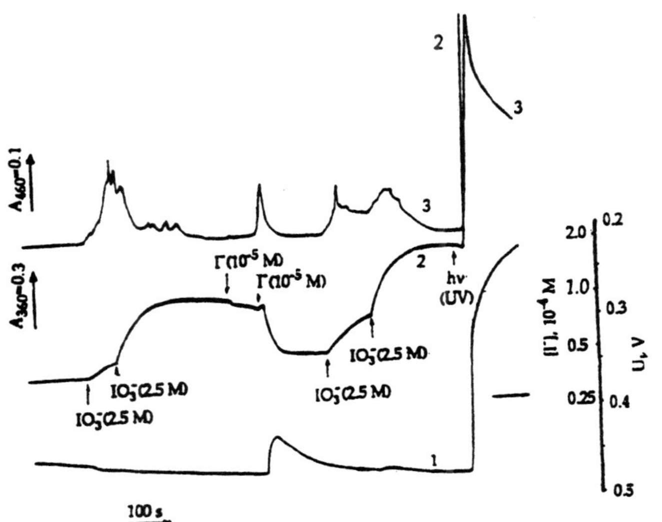

Figure 6. Production of state (**I**′) in the Briggs–Rausher system by adding iodate to a 0.1 M H_2SO_4/0.61 M H_2O_2/0.06 M MA/ 0.035 M Mn^{2+} mixture, and the kinetic phase transition (**I**′) → (**II**) upon addition of iodide to or UV illumination of state (**I**′). The recorded parameters are the iodide electrode potential U_{I^-}, A_{360}, and A_{460} versus time.

The concentrations of I_2 and I^- are low. Addition of iodide or illumination result in the breakdown of R_xI (Figure 6). Under weak stirring, spontaneous breakdown of R_xI is also possible. Assuming that all iodate converts to R_xI, the increment in A_{360} yields for R_xI an e_{360} of 65 M^{-1} cm^{-1}. State (**I'**) differs from (**I**) only in that because of the different initial reagent concentrations (MA, IMA, R_xI, H_2O_2) autocatalysis is markedly suppressed during the (**I'**) → (**II**) transition and the latter proceeds in a quasi-steady-state mode, i.e., more slowly than the (**I**) → (**II**) transition (which is typical of unbranching chain reactions). In the process, reactions (6), (7) and (13) consuming iodine and iodide prevail over the reactions of R_xI decomposition that produce I_2 and I^-. As a result, the registered values of $[I_2]$ and $[I^-]$ during the (**I'**) → (**II**) transition are significantly lower than during the (**I**) → (**II**) transition, while the drop in A_{360} becomes clearly observable and is characterized by a certain maximal rate V_{max} as in the case of curve % in Figure 3.

The existence of an induction period τ in autocatalytic formation of I_2 and I^-, and the autocatalytic and quasi-steady-state modes of R_xI breakdown are indicative of two possible regimes of a chain reaction — branching and unbranching, the switch-over between them depending on the concentration of some inhibitor. It is known [15] that threshold effects are possible in such systems. The existence of threshold phenomena in the (**I**) → (**II**) transition is also confirmed by the experiments on photoinduced transition [11] which will be considered in more detail in the next work.

Cook [10] has supposed that IMA is the source of I_2 in the Briggs–Rausher system. However, we have observed a number of facts testifying that there should be and there is another source of iodine, i.e., R_xI is not IMA. The following arguments can be produced in favor of this assertion.

First, it should be emphasized (though it is practically obvious) that the increase in absorbance at 360 nm is due to accumulation of an organic compound and not . Indeed, with a simultaneous decline in $[I_2]$ and $[I^-]$ by the end of and immediately after the cessation of oscillations, the triiodide concentration should decrease in accordance with Equation (10), whereas A_{360} begins to increase, and at a rate substantially exceeding the average one during oscillations (Figure 2). Besides, the contribution of triiodide to A_{360} at $[I_2] < 10^{-6}$ M and $[I^-] < 4 \cdot 10^{-6}$ M cannot exceed $2 \cdot 10^{-4}$.

Second, if IMA is mainly formed in reaction (7), the rate of the increase in [IMA] cannot be greater than $k_7[MA]\{I_2\}$. Therefore, using the $[I_2]$ values determined from the curve in Figure 2 one can find that after the end of oscillations the $d A_{360}/dt$ should not exceed $7.2 \cdot 10^{-4}$ sec^{-1}, whereas the experimentally observed rate is an order of magnitude higher: curve 4 in Figure 2 yields a value of about $7 \cdot 10^{-3}$ sec^{-1}.

Third, if one assumes that the increment in absorbance at 360 nm is due to IMA alone, from the A_{360}^{max} and ϵ_{360}^{IMA} it follows that $[IMA_{max}] \approx$ 0.06 M, i.e., exceeds several-fold the initial concentration of iodate (0.0225 M) which is the source of iodine in the Briggs–Rausher system.

Fourth, it can be seen in Figures 2 and 3 that A_{360} reaches a maximum and then slowly declines, indicating that the R_xI compound formed is unstable. AT the same time, IMA produced in the I^- $/H_2O_2/MA/H_2SO_4$ mixture with further addition of Mn^{2+} or in the Mn^{2+}/I^- $/H_2O_2/MA/H_2SO_4$ mixture is quite a stable compounds.

Fifth, the $MA/IMA/H_2O_2/Mn^{2+}/H_2SO_4$ mixture is not simply stable but virtually insensitive to light (the reagent concentrations in the mixture have been chosen to approximate those in state (**I**) of the Briggs–Rausher system). On the other hand, state (**I**), which apart from R_xI is the same mixture, is rather photosensitive.

All this provides weighty evidence for the existence in state (**I**) and the decisive role of a certain compound R_xI distinct from IMA.

4. Putative Mechanism of the Autocatalytic Step

We do not know the exact mechanism of the autocatalytic step described here. If R_xI is diiodomalonic acid I_2MA which can be formed in the reaction

$$I_2 + IMA \rightarrow I^- + I_2MA + H^- \tag{14}$$

(or diiodoacetic acid), two versions can be envisaged for autocatalytic formation of I_2 and I^-. The first version comprises the following reactions:

$$H^+ + I_2MA + I^- \rightarrow IMA + I_2 \tag{15}$$

$$I_2 + HO_2^{\cdot} \rightarrow O_2 + I^- + I^{\cdot} + H^+ \tag{16}$$

$$I_2MA + I^{\cdot} \rightarrow I^2 + IMA^{\cdot} \tag{17}$$

$$H_2O + IMA^{\cdot} \rightarrow I^- + H^+ R_y^{\cdot} \tag{18}$$

Summation of reactions (15 to (18) gives

$$HO_2^{\cdot} + 2I_2MA + I^- \rightarrow O_2 + I_2 + 2I^- + IMA + R_y^{\cdot} + H^+ \tag{19}$$

where

$$R_y^{\cdot} \equiv HOOC-\overset{\overset{\large O^{\cdot}}{|}}{C}H-COOH$$

The second version comprises the reactions

$$H^+ + I_2MA + I^- \rightarrow IMA + I_2 \tag{15}$$

$$I_2 + MA \rightarrow IMA + I^- + H^- \tag{7}$$

$$IMA + MA^{\cdot} \rightarrow MA + IMA^{\cdot} \tag{20}$$

$$H_2O + IMA^{\cdot} \rightarrow I^- + H^+ + R_y^{\cdot} \tag{18}$$

with the sum reaction

$$MA^{\cdot} + I_2MA + I^- \rightarrow 2I^- + IMA + R_y^{\cdot} + H^+ \tag{21}$$

Combinations of these versions are also possible. Reactions (19) and (21) explain not only the autocatalytic rise in I^- and I_2 but also the reason why the **(I)** → **(II)** transition is initiated by addition of iodide. Reactions (15) and (7) provide an unbranching chain mechanism for the breakdown of I_2MA. The balance between the iodide-producing (18, 16, 14 and 17) and the iodide-consuming reactions (15, 13, 12, 9 and 6) may regulate the value of the threshold for the switchover from the unbranching chain mechanism to the branching, autocatalytic one. At the same time, competition between the key reaction of autocatalysis (15) and reactions (6), (9) and partly (13) which inhibit reaction (15) determines the threshold for the onset of any chain process.

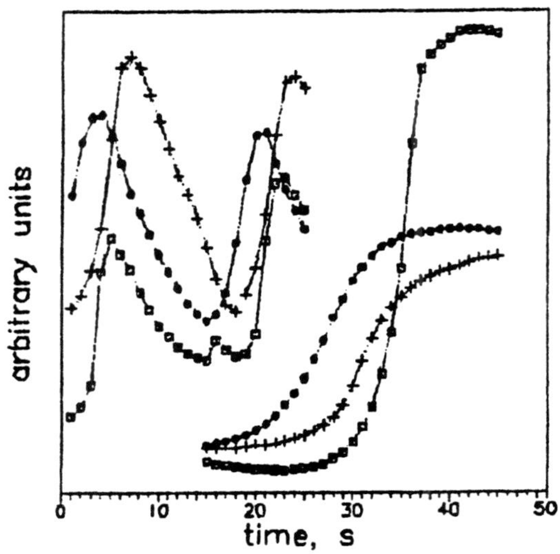

Figure 7. Phase relations between the absorbance at 460 nm (iodine concentration, •·), the iodide-selective electrode potential (iodide concentration, +), and absorbance at 360 nm (sum concentration of triiodide and malonic acid iododerivatives, o) in the Briggs–Rausher reaction in the oscillatory mode and during the (**I**) → (**II**) transition evoked by addition of iodide. Temp. 25° C, $[I^-]_{add} = 5 \cdot 10^{-5}$ M; the moment of iodide addition to state (**I**) (600 sec after the onset of the reaction) is aligned with the beginning of the rise in $[I_2]$ in the oscillatory mode.

Each of the proposed versions involves radicals. These may arise in initiation reactions like

$$I_2MA \xrightarrow{H_2O_2,\, Mn^{2+}} I_2 + IR^{\cdot}_y + HO^{\cdot}_2 \tag{22}$$

or by thermal decomposition of hydroperoxy derivatives of malonic or iodomalonic acid:

$$IMAOOH \rightarrow IR^{\cdot}_y + OH^{\cdot} \tag{23}$$

or in the reaction with Mn^{2+}:

$$MAOOH + Mn^{2+} \rightarrow R_y^{\cdot} + MnOH^{2+} \quad (24)$$

$$MnOH^{2+} + H_2O_2 \rightarrow Mn^{2+} + HO_2^{\cdot} + H_2O \quad (25)$$

Besides the initiation reactions, the system naturally includes a complicated sequence of radical interconversions, and stages of chain termination like

$$2I^{\cdot} \rightarrow I_2$$

$$MA^{\cdot} + I^{\cdot} \rightarrow IMA$$

$$HO_2^{\cdot} + HO_2^{\cdot} \rightarrow O_2 + H_2O_2$$

In conclusion, it appears worthwhile to compare the phase relations between I_2, I^- and I_3^- during the (**I**) → (**II**) transition with the corresponding relations in the oscillatory mode of the Briggs–Rausher reaction. Figure 7 shows that in both cases iodide starts to be autocatalytically produced with a small lag relative to iodine. Besides, on the time curves for A_{360} in both cases there is a small drop immediately after the start of the rise in $[I_2]$. As already mentioned, in the (**I**) → (**II**) transition this drop is due to the breakdown of R_xI. All this gives grounds for putting forward, in addition to the skeleton model available, a novel mechanism of the Briggs–Rausher reaction involving an autocatalytic step of I_2 and I^- formation owing to breakdown of R_xI, with Mn^{2+}, H_2O_2, and IO_3^- being necessary to form R_xI and to generate radicals.

Acknowledgment

This work was supported through the project "Microcapsules" coordinated by Prof. M.V. Alfimov.

5. References

1. **T.S. Briggs** and **W.C. Rausher.** *J. Chem. Educ.,* Vol. 50, 1973, p. 496.
2. **R.M. Noyes** and **S.D. Furrow.** *J. Am. Chem. Soc.,* Vol. 104, 1982, p. 45.
3. **P. DeKepper** and **I.R. Epstein.** *J. Am. Chem. Soc.,* Vol. 104, 1982, p. 49.
4. **S.D. Furrow.** In: *Oscillations and Traveling Waves in Chemical Systems* (R.J. Field and M. Burger, eds.), New York: Wiley-Interscience, 1985, p. 171.
5. **D. Edelson.** *J. Phys. Chem.,* Vol. 87, 1983, p. 1204.
6. **P. DeKepper** and **J. Boissonade.** In: *Oscillations and Traveling Waves in Chemical Systems* (R.J. Field and M. Burger, eds.) New York: Wiley-Interscience, 1985, p. 223.
7. **Z. Noszticzius, E. Noszticzius** and **Z.A. Schelly,** *J. Am. Chem. Soc.,* Vol. 104, 1982, p. 6194.
8. **V.K. Vanag, A.N. Kuznetsov** and **L.A. Piruzyan.** *Biofizika,* Vol. 28, No. 1, 1983, p. 18 (in Russian).
9. **S.D. Furrow.** *J. Phys. Chem.,* Vol. 93, 1989, p. 2817.
10. **D.O. Cook.** *Reaction Kinetics and Catalytic Letters,* Vol. 27, No. 2, 1985, p. 379.
11. **V.K. Vanag, A.A. Schelypin** and **M.V. Alfimov.** *Doklady Akad. Nauk SSSR,* Vol. 316, No. 2, 1991, p. 388 (in Russian).
12. **V.A. Vavilin, A.M. Zhabotinskii** and **A.N. Zaikin.** *Zhurnal Fizicheskoi Khimii,* Vol. 32, No. 12, 1968, p. 3091 (in Russian).
13. **L.P. Tikhonova, A.S. Kovalenko** and **I.F. Labunskaya.** *Zhurnal Neorganicheskoi Khimii,* Vol. 33, No. 3, 1988, p. 649 (in Russian).
14. **S.D. Furrow.** *J. Phys. Chem.,* Vol. 85, No. 14, 1981, p. 2026.
15. **N.N. Semenov.** *Tsepnye Reaktsii* (Chain Reactions), Leningrad: ONTI, 1934, 360 pp. (in Russian).

CHEMICAL MODEL OF OXYGENASES. CU(I) COMPLEXES AS CATALYSTS OF ISOPROPANOL OXIDATION BY DIOXYGEN IN ALKALINE MEDIUM

A. M. Sakharov and I. P. Skibida
Institute of Chemical Physics,
The Russia Academy of Sciences, Moscow

1. SUMMARY

Isopropanol oxidation was investigated to establish the mechanism of oxidation of secondary alcohols by dioxygen in the presence of o-phenanthroline complexes of copper in alkaline solutions.

At 20° C and a base-to-catalyst ratio of 1 to 2, oxidation proceeds at a high rate close to that of the enzymatic reaction with a turnover number of 1–2 s^{-1}.

The unique oxidation product is acetone; the selectivity of its formation in terms of oxygen consumed approaches 100%.

It was shown that the catalytically active species is Cu(I) bis(o-phenoanthrolinate) (Cu^+L_2) and the base acts as a co-catalyst. Oxidation proceeds by a two-electron single-step mechanism with the intermediate formation of the ternary complex $[O_2...Cu^+L_2...^-OC_3H_7]$. An essential step of the reaction is hydrogen transfer. The magnitude of the kinetic isotope effect. $k_H/k_D = 2$ is consistent with the assumption that this step is the rate limiting one.

The designing of chemical catalytic systems simulating the activity of oxidases and oxygenases is one of the most attractive objectives for the chemistry of catalysis by metal complexes.

Numerous studies in this field have provided interesting results on the modelling of oxygenases and oxidases using hydrogen peroxide and other single oxygen donors as oxidants [1]. However up to day there is no examples of catalytic systems resembles emzymatic ones in the efficiency of oxygenation or oxidative dehydrogenation using dioxygen as an oxidant. The biomimetic systems proposed thus far allow the oxygen-mediated oxidation of only readily oxidizable substrates, such as phenols, ascorbate anions and hydroxyketones [2–8].

As far as these compounds exhibit high redox activities and readily react not only with O_2 but also with its reduced forms as well as with various metal complexes used as catalysts and their transformation products, the mechanisms proposed are largely hypothetical.

A high catalytic activity of copper complexes with p-phenanthroline and bipyridine in the oxidation of methanol and methylethylketone by dioxygen in alkaline media has been firstly demonstrated in [9]. Our detailed investigations of the oxidation mechanism and subsequent optimization of this catalytic system have allowed the oxidation rates of primary alcohols to be increased by 1–2 orders of magnitude and extraordinarily high selectivities and yields to be attained in many cases.

It has been shown that the strong dependence of catalytic activity, selectivity and oxidation route of primary alcohols and ketones on reaction conditions is largely due to a change in the mechanism of oxygen activation, namely to the switch from a sequence of one-electron steps to single-step two-electron oxidation in the coordination sphere of the catalyst.

Under optimum conditions the activity of copper-containing catalysts in the oxidation of primary alcohols and ketones approaches that of enzymatic systems.

Studies on the oxidation of primary alcohols to aldehydes in the presence of copper-phenanthroline complexes in non-aqueous media have suggested the following reaction mechanism [10]:

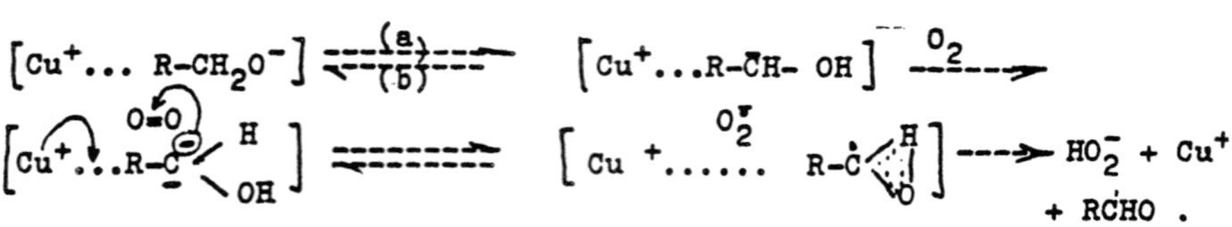

An essential step in the proposed mechanism is proton transfer with the formation of the carbanion $R\bar{C}HOH$ (step a).

Secondary alcohols are not oxidized at appreciable rates under conditions optimal for the oxidation of primary ones and can, therefore, be used as neutral solvents.

In the present study conditions for the oxidation of secondary alcohols by dioxygen in non-aqueous media in the presence of alkali and copper complexes with o-phenanthroline were investigated using isopropanol and its mixtures with primary alcohols in order to further clarify the catalytic mechanism and find the reasons for highly selective oxidation of primary alcohols in the presence of secondary ones.

2. EXPERIMENTAL

The oxidation of H_8-isopropanol was usually carried out in the absence of solvents. Primary alcohols were added in some particular experiments. D_7-isopropanol ($CD_3CD(OH)CD_3$) was oxidized in acetonitrile.

Copper complexes of o-phenanthroline, which were used as catalysts, were prepared by dissolving CuCl or $CuCl_2 2H_2O$ and o-phenanthroline in isopropanol in a molar ratio of 1:4. In the case of D_7-isopropanol oxidation the components were dissolved in acetonitrile at the ratio of [CuCl]:[o-phen] = 1:15. Excess of o-phenanthroline was used to stabilize the phenanthroline complexes of Cu(I) in alkaline solutions. In neutral medium, CuCl gives rise to bis- and $CuCl_2$ to tris-phenanthroline complexes, $[Cu(o\text{-}phen)_2]+$ and $[Cu(o\text{-}phen)_3]^{2+}$, respectively. The latter are unstable in the presence of alkali and decompose to give free o-phenanthroline and hydroxyphenanthroline complexes of varying composition [12].

The alkali (0.5–1.0 M solution of sodium or potassium hydroxide in isopropanol) was injected into the continuously stirred reaction mixture from a syringe. In some experiments, granular alkali was placed in the solution before starting the reaction.

Molecular oxygen was used as an oxidant.

The reaction rate was measured as the rate of oxygen uptake or of the accumulation rate of acetone, the main reaction product.

Acetone was quantified by GLC on a 2 m x 0.2 ÷ 0.31 mm Rysorb BLK column impregnated with 10% polyethylene glycol adipate. A flame ionization detector was used.

The concentration of bis-phenanthroline complexes of Cu(I) was measured by means of UV-VIS spctroscopy at a wavelength of 430 nm using an extinction coefficient of $\varepsilon = 7040$ [13].

3. Results and Discussion

3.1. Kinetics of Catalytic Oxidation of Isopropanol

The effect of alkali on the rate of isopropanol oxidation was studied using both $[Cu(o\text{-}phen)_2]^+$ complexes and phenanthrolinates of bivalent copper as the initial catalyst forms.

As is the case with the oxidation of primary alcohols [10], isopropanol oxidation in the presence of copper phenanthrolinates does not start until a base is added to the solution.

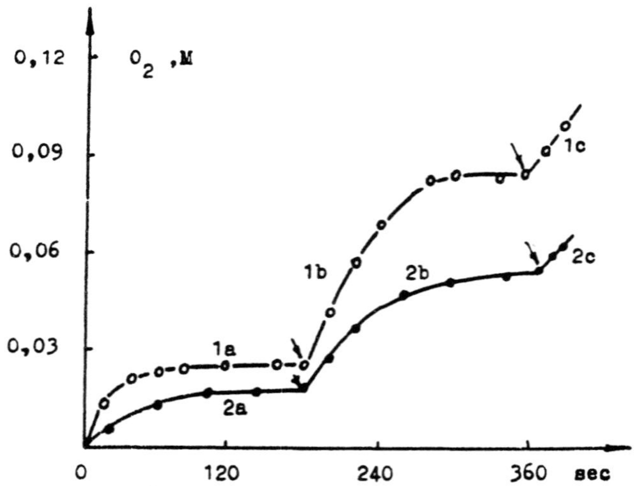

Figure 1. Kinetic curves of oxygen uptake during isopropanol oxidation at 25° C. Solution volume is 10 ml.
1–$[KOH]_o$ = 50 mM, $[Cu(o\text{-}phen)_2]^+_o$ = 5 mM;
2–$[KOH]_o$ = 50 mM, $[Cu(o\text{-}phen)_2]^+_o$ = 5 mM;
(the arrow indicates the addition of a KOH aliquote to a concentration of 5 mM).
3–$[KOH]_o$ = 10 mM (in the granular form), $[Cu(o\text{-}phen)_3]^{2+}$ = 5 mM.

The addition of equimolar quantity of KOH to isopropanol solution containing $[Cu(o\text{-}phen)_2]^+$ results in an extremely high rate of oxygen uptake (Figure 1, curve 1). The main oxidation product under such conditions is acetone. The molar ratio of oxygen consumed to acetone formed is close to 2, which means, in accordance with the stoichiometric equation

$$CH_3CH(OH)CH_3 + 1/2\ O_2 = CH_3C(O)CH_3 + H_2O\ ,$$

that the selectivity of acetone formation in terms of oxygen consumed is close to 100%.

Isopropanol oxidation ceases as soon as the alkali is entirely used up due to further oxidation of acetone to acids and, probably, to side reactions involved slow direct oxidation of the alcoholate ions $(CH_3)_2CHO^-$ to acids.

The repeated addition of alkali (Figure 1, curve 1) completely restores the activity of the catalyst and the high oxidation rate.

The base acts as a co-catalyst since its consumption does not exceed 1 mole per 100 moles of acetone.

The activity of Cu(I) complexes in the oxidation of isopropanol to acetone significantly exceeds that of most catalysts known thus far and approaches the activity of enzymes catalyzing the oxidation of organic compounds by dioxygen. The turnover number was estimated as 1–2 s^{-1}. The catalyst remains active even after several thousands of catalytic cycles.

Increasing the KOH (or NaOH) content in isopropanol up to $[OH^-]_o/[Cu^{2+}]_o$ ratio of about 10 results in almost instantaneous hydrolysis of the catalyst regardless of its valent state in the initial form. This is accompanied by the formation of a brown precipitate containing unidentified copper compounds and complete cessation of oxygen uptake (Figure 1, curve 2).

Unlike isopropanol oxidation, the oxidation rate of primary alcohols in the presence of phenanthroline complexes of Cu(I) significantly increases with the alkali content up to $[OH^-]_o/[Cu^{2+}]_o$ ratios of 5–10 and decreases due to the hydrolysis of the complexes only in more alkaline media [10]. No brown precipitates are usually formed in this case pointing to a different mechanism of catalyst deactivation.

In the presence of complexes of bivalent copper with o-phenathroline as the initial catalyst form, the situation is somewhat more complicated. If the quantity of KOH solution added to a solution of $[Cu(o\text{-}phen)_3]^{2+}$ in isopropanol is equimolar to that of the catalyst then the latter remains divalent and no oxygen uptake is observed. In this case divalent copper complexes appear to act as monobasic Lewis acids to fully neutralize the medium.

At a twofold excess of the alkali the reaction starts with a short autoacceleration period during which Cu(II) transition to Cu(I) takes place. The most distinct induction period is observed when the alkali is added in the granular form since Cu(II) to Cu(I) transition is limited in this case by slow dissolution of the alkali in isopropanol. The rate of oxygen uptake in the course of the reaction is practically proportional to the concentration of monovalent copper.

The maximum rates of isopropanol oxidation with the catalyst being initially in the Cu(I) and Cu(II) forms are virtually identical. In both cases the concentration of Cu(I) complexes comes to steady state, the catalytically active complexes $[Cu(o\text{-}phen)_2]^+$ accounting for 30 to 50% of the total amount of the catalyst added.

The steady state level of $[Cu(o\text{-}phen)_2]^+$ depends on the base concentration, and decreasing the latter would eventually result in the complete catalyst transition to the monovalent form. However, this would not increase the oxidation rate since the concentration of the active catalyst form grows simultaneously with decreasing the level of alcoholate ions. In strongly alkaline media, where the concentration of alcoholate ions is sufficiently high, monovalent copper is practically not formed and no oxygen uptake is observed.

The results obtained allow a conclusion that isopropanol oxidation to acetone is catalyzed by monovalent copper complexes as is the case with the oxidation of primary alcohols to aldehydes [10].

3.2. Oxidation of Isopropanol in the Presence of Secondary and Primary Alcohols

Cyclohexanol, secondary butanol and other secondary alcohols have virtually no effect on the rate of oxygen consumption during the oxidation of isopropanol, but isopropanol conversion and acetone yield in terms of alkali added significantly decrease. This is due, apparently, to the fact that the oxidation rates of cyclohexanol and secondary butanol under the conditions of catalysis are similar to that of isopropanol oxidation. However unlike acetone, which may be considered as the end product, methylethylketone and cyclohexanone undergo further rapid oxidation to acids [11,14] resulting in the neutralization of the medium and cessation of the reaction.

Primary aliphatic alcohols decrease significantly both the oxygen consumption rate and $\Delta[O_2]/\Delta[KOH]$ ratio. The addition of 5% primary alcohols to isopropanol is sufficient to decrease $\Delta[O_2]/\Delta[KOH]$ and $\Delta[CH_3COCH_3]/\Delta[KOH]$ ratios more than 10-fold (Table 1).

That reaction in the presence of primary alcohols stops at low conversions is due to the rapid neutralization of the medium resulting from high rates of RCH_2O^- oxidation to aldehydes, the latter undergoing aldol condensation followed by the oxidation of the aldols to acids.

Table 1. Oxidation of isopropanol in the presence of 5% secondary and primary alcohols. $[Cu(o\text{-}phen)_2]^+{}_0$ = 5 mM, $[KOH]_0$ = 3 mM. Solution volume is 10 ml, T = 30° C.

Alcohol	$W(O_2)$ max 10^3 M/s	Δ $[O_2]$ nM	[Acetone] nM	$\Delta[O_2]/\Delta[KOH]$	Δ[Acetone]/Δ[KOH]
None	1.5	150	290	50.0	96.7
Secondary butanol	1.3	70	130	23.3	43.3
Cyclohexnol	1.3	50	190	16.7	30.0
Methanol	0.4	14	20	4.7	6.7
Ethanol	0.5	18	30	6.0	10.0
n-Propanol	0.4	12	19	4.0	6.3
n-Butanol	0.35	13	20	4.3	6.7

Table 2. Effect of n-butanol on rate of isopropanol oxidation and proportion of products.
$[Cu(0\text{-}phen)_3]^{2+}{}_0$ = 5 mM, $[KOH]_0$ = 30 mM (4 consecutive additions, 7.5 mM each). Solution volume is 10 ml. T = 30∞ C.

n-Butanol content in its mixture with isopropanol, vol. %	$W(O_2)_{max}10^3$ M/s	$\Delta[O_2]$ mM	$[CH_3COCH_3]$ mM	$[C_3H_7CHO]$ mM
0	1.1	550	1050	—
5	1.0	120	160	5
10	0.95	95	120	10
25	0.75	65	80	20
50	0.45	60	35	35
75	0.19	60	5	70
90	0.12	65	0	90
100	0.12	70	—	100

As demonstrated earlier by us [10] the formation of adducts between the alcoholate ions RCH_2O^- and the catalyst is essential for the oxidation of primary alcohols. The oxidation of secondary alcohols involves, apparently, the formation of similar adducts. The highly efficient inhibition of isopropanol oxidation is likely to be due to the efficient solubilization of Cu(I) complexes by primary alcohols and a higher stability of catalyst adducts with the anionic forms of primary alcohols as compared to adducts with isopropanol. This results in competitive inhibition.

The equilibrium constant for the formation of adducts with univalent copper ions appears to be higher for benzyl alcohol than for aliphatic primary ones. This is seen from the fact that $C_6H_5CH_2OH$ can be oxidized to benzaldehyde at benzyl alcohol contents in isopropanol below 10% (v/v), with no traces of acetone being detectable in the solution [10].

A higher efficiency of interaction between primary alcohols and the catalyst, at least as regards the formation of catalyst adduct with the anionic form of the substrate, is further demonstrated by experiments on the oxidation of isopropanol-n-butanol mixtures of varying composition. In these experiments, KOH was added portionwise which allowed greater conversions to be attained.

About 15% of the catalyst in the starting cuprous chloride solution exists in the Cu(II) form due to the partial oxidation of CuCl. Therefore, the alkali added is partly spent on the neutralization of Cu(II), which, in turn, diminishes the oxidation rate and conversion of isopropanol.

With portionwise alkali addition, at the end of the first step all catalyst is in the Cu(I) form and all subsequent KOH aliquotes are entirely spent on the neutralization of acids.

As seen from Table 2, the dilution of isopropanol with primary alcohol decreases the rate of oxygen uptake and acetone yield. At isopropanol contents of 50% and 25% in its mixtures with $C_3H_7CH_2OH$ the yield of acetone decreases by a factor of 30 and 200, respectively, as compared to the bulk oxidation of isopropanol.

The complexes of Cu(I) with the alcoholate ions from primary alcohols are hydrolyzed much more slowly than complexes with secondary alcohols. This permits the oxidation of the former to be conducted in more alkaline media, where the concentration of the anionic form is much higher than in the case of isopropanol oxidation. The maximum consumption rate of primary alcohols under such conditions is close to that of isopropanol oxidation in slightly alkaline media, which demonstrates the $Cu(I)RO^-$ adducts of secondary alcohols as being more reactive towards dioxygen.

3.3. Oxidation of 1,2,3-D_7-Isopropanol

The oxidation of 1,2,3,-D_7-isopropanol was investigated in order to evaluate the magnitude of the kinetic isotope effect. The experiments were carried out at a catalyst concentration of $3 \cdot 10^{-3}$ M with the portionwise addition of 0.9 M KOH to the reaction mixture. The aliquotes added accounted for as much as 2% of the initial solution volume and had no appreciable effect on the reaction rate.

The kinetic isotope effect was evaluated as the ratio of oxygen consumption rates during the oxidation of isopropanol and 1,2,3-D_7-isopropanol after the second addition of 0.02 ml of 0.9 M KOH to 10 ml of the reaction mixture.

The $[Cu(I)]_o$ concentration in the starting solution was $2.45 \cdot 10^{-3}$. Before the addition of the second aliquote of alkali $[Cu(I)]_o$ was equal to $3 \cdot 10^{-3}$ M, i.g. that all catalyst was in the form of $[Cu(o\text{-}phen)_2]^+$ complexes. In the course of the reaction, the concentration of univalent copper varies within 10% in oxidation both isopropanol and D_7-isopropanol (Table 3).

The kinetic curves of oxygen uptake during the oxidation of i-C_3H_7OH and i-C_3D_7OH are presented in Figure 2.

Table 3. Content of $[Cu(o\text{-}phen)_2]^+$ in solution during oxidation of i-C_3H_7OH and i-C_3D_7OH after the second addition of alkali (conditions of oxidation and measuring times are the same as in Figure 2).

Time, s	$[Cu(o\text{-}phen)_2]^+ \cdot 10^3$	
	i-C_3H_7OH	i-C_3D_7OH
0	2.45	2.45
180	3.0	3.0
200	2.75	—
230	—	—
240	2.8	—
280	3.0	—
360	3.0	3.0

Table 4. i-C_3H_7OH and i-C_3D_7OH oxidation.

Substrate	$\Delta [O_2]$, M	[Acetone], M	[Acids], M
isopropanol	0.18	0.34	0.005
D_7-isopropanol	0.10	0.165*	0.005

*The concentration of CD_3COCD_3 was determined using non-deuterated acetone as a standard.

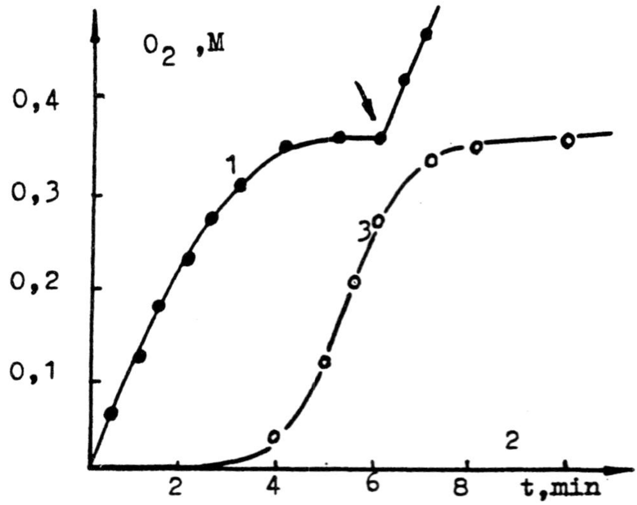

Figure 2. Kinetic curves of oxygen uptake during the oxidation of 30% solutions of i–C_3H_7OH (1) and i–C_3D_7OH (2) in acetonitrile. The catalyst concentration is $3{\cdot}10^{-3}$ M. T = 20° C. The arrows indicate the addition of 0.02 ml of 0.9 M KOH in alcohol.

The rates of oxygen uptake were measured 10 seconds after the addition of the second KOH aliquote in order to avoid the effect of mixing. The ratio of the oxygen uptake rates gives the following value of the kinetic isotope effect

$$\frac{W^{H}_{O_2}}{W^{D}_{O_2}} = \frac{1.35 \cdot 10^{-3}\, M/s}{0.65 \cdot 10^{-3}\, M/s} = 2.1.$$

The ratio remains close to 2 throughout the course of the reaction.

As seen from Figure 2, conversion (the amount of oxygen consumed) is higher with i-C_3D_7OH than with H_8-isopropanol (see Table 4). The main oxidation product in both cases is acetone. A small quantity of acids is formed as by-products which neutralize the medium thus arresting oxidation. With isopropanol, 0.5 mg. eqv. of acids is formed after 18 moles of oxygen being consumed and 34 moles of acetone formed. With deuterated isopropanol, the same amount of acids is formed after the consumption of as little as 10 moles of oxygen.

Although deuteration reduces the oxygen uptake rate almost by half, it obviously does not affect significantly (probably somewhat increases) the side reactions of acetone oxidation to acids. This appears to account for the reduced total amount of oxygen consumed at the same quantity of alkali added.

3.4. MECHANISM OF ISOPROPANOL OXIDATION

The data obtained demonstrate that the coordinatively saturated complexes $[Cu(o\text{-}phen)_2]^+$ are those species responsible for the catalytic activity of the catalyst. Their turnover number of $1–2s^{-1}$ is, as pointed out above, close to the activity of a series of enzymes catalyzing the oxidation of organic substances by dioxygen.

Another postulated active species is the coordinatively unsaturated complex $[Cu(o\text{-}phen)]^+$ formed on the dissociation of the complex $[Cu(o\text{-}phen)_2]^+$. If this were true then, in view of an extremely high stability of Cu(I) bis-phenanthrolinates [13] and, consequently, low steady-state concentration of $[Cu(o\text{-}phen)_2]^+$ in the presence of excess of free phenanthroline, the turnover number of this species would be 10^5–$10^6 s^{-1}$, which is quite unlikely for reactions of the kinetically rather inert dioxygen.

As with the oxidation of primary alcohols, the first step in the oxidation of isopropanol appears to involve the formation of an adduct between $[Cu(o\text{-}phen)_2]^+$ and the anionic form of the substrate (R_2CHO^-). The adduct is formed almost immediately after the addition of the alkali to the system. In strongly alkaline media such an adduct undergoes further conversions (probably with the participation of a second alcoholate ion), and catalytically inactive brown precipitate is formed within several seconds. The formation of inactive catalyst forms is accelerated with increasing the electron-donor ability of the alcoholate ions as is the case, for instance, when isopropanol is diluted by such aprotic solvents as acetonitrile, DMF, dioxane, etc. In this case the hydrolysis and deactivation of the catalyst take place at lower concentrations of the alkali.

By its kinetic behaviour the system under study is quite similar to galactose oxidase [2] whose active center is known to contain a Cu^+ ion responsible for the oxidation of primary hydroxyl to carbonyl groups.

With Cu(I) phenanthrolinates, the adduct [Cu(I)(o-phen)$_2$...R_2CHO^-] appears capable of readily reacting with O_2 as well. In the absence of alcoholate ions, $[Cu(I)(o\text{-}phen)_2]^+$ is practically incapable of reacting with O_2. It is possible, however, that the formation of the adduct [Cu(I)(o-phen)$_2$...R_2CHO^-] sharply increases the donor ability of Cu(I) complexes, especially as Cu(I) in such a complex is pentacoordinate that is, coordinatively oversaturated.

It is usually assumed that the first step in the reactions catalyzed by galactose oxidases involves the interaction of Cu(I) L_2 with O_2 followed by electron transfer to give O_2^- and a Cu(II) complex [2,3]. It is postulated that the active species is a Cu(III) complex which oxidizes an alkoxy to a carboxy group:

$$[R_2CHO^- \ldots Cu(I) \ldots O_2] \leftrightarrow [\,R_2CHO^- \ldots Cu(II) \ldots O_2^-] \leftrightarrow$$

$$\leftrightarrow [R_2CHO^- \ldots Cu(III) \ldots O_2^{=} \leftrightarrow R_2C = O + Cu(I) + HO_2^-$$

It is assumed [2,3] that reactions with the participation of Cu(III) complexes are a very common occurrence, difficulties with the identification of this species being due solely to a high reactivity and low steady-state concentration of Cu(III) complexes in solution which result, in particular, from high rates of ligand oxidation by Cu(III).

In the catalytic oxidation of isopropanol, like in the oxidation of primary alcohols in the presence of [Cu(I)o-phen)$_2$], the ligand (o-phenanthroline) remains intact even after several thousands of catalytic cycles. The high stability of the ligand in the system under consideration casts doubt on the possibility of O_2^- formation as a reactive intermediate since superoxide is known to react readily with o-phenanthroline liganded by a transition metal to destroy it [15].

We believe that Cu(I)-liganded alcoholate ions may be oxidized by a concerted mechanism without the intermediate formation of superoxide and Cu(III) complexes. One can suggest that isopropanol oxidation, which represents formally hydride ion transfer from an alcoholate ion to O_2, involves the simultaneous transfer of two electrons and a proton to an oxygen molecule:

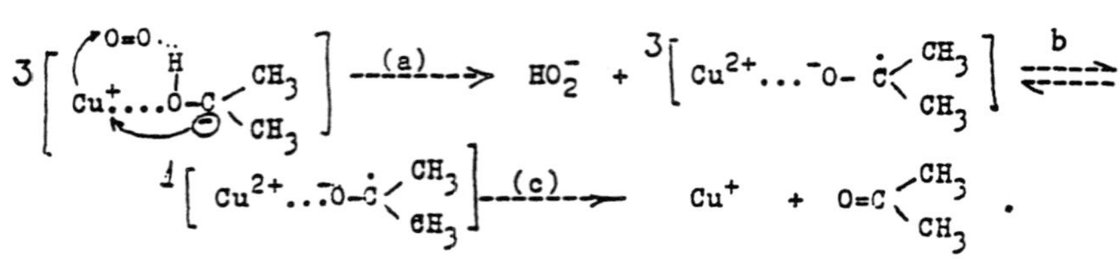

Our investigations demonstrated that hydrogen peroxide formed during isopropanol oxidation undergoes rapid disproportionation to oxygen and water and takes no part in the formation of reaction products.

The proposed scheme of oxygen activation is in a good agreement with the magnitude of the kinetic isotope effect for i-C_3D_7OH oxidation ($k_H/k_D = 2$) which is characteristic of processes with hydrogen transfer as the rate-limiting step (step a) [16].

5. References

1. **B. Meunier**, Bull. Soc. Chim. Fr., 1986, pp. 578–587; Gazzetta Chim. Ital., Vol. 118, 1988, pp. 485–493.
2. **G. A. Hamilton, P. K. Adolf, J. de Jersey, G. S. DuBois, G. R. Dyrkaez, and R. D. Libby**, J. Amer. Chem. Soc., Vol. 100, No. 6, 1978, pp. 1899–1912.
3. **L. D. Kwiatkowski, M. Adelman, R. Pannelly, and D. J. Kosman, J. Inorg.**, Chem., Vol. 14, 1981, pp. 209–22.
4. **P. Capdeville, P. Audebert, and M. Maumy**, Tetrahedron Lett., Vol. 25, No. 39, 1984, pp. 4397–4402.
5. B. S. Tovrog, E. E. Diamant, and F. Mares, J. Amer. Chem. Soc., Vol. 101 Chem. Communs., No. 17, 1979, pp. 5067–5069.
6. **M. Mumakata, S. Nishibayashi, and H. Sakamoto**, J. Chem. Soc., V. 101, Chem. Communs., No. 5, 1980, pp. 220.
7. **J. J. Discoll and D. J. Cosman**, J. Amer. Chem. Soc., Vol. 109, No. 6, 1987, pp. 1769–1777.
8. **K. B. Wiberg and W. G. Nigh**, J. Amer. Chem. Soc., Vol. 87, No. 17, 1965, pp. 3849–3855.
9. **W. Brackman and C. J. Gaasbeek**, Rec. trav. chim. Pays-Bas., Vol. 85, No. 2, pp. 242–256.
10. **A. M. Sakharov, and I. P. Skibada**, J. Molec. Catal., Vol. 48, No. ", pp. 157–174.
11. **A. M. Sakharov, Ja. V. Muminov, and I. P. Skibida**, Kinetica i Kalatiz, Vol. 30, No. 1, 1989, pp. 112–120 (in Russian).
12. **A. M. Sakharov and I. P. Skibada**, Dokl. AN SSSR, Vol. 251, No. 1, pp. 155–159 (in Russian).
13. **R. G. P. Williams**, J. Chem. Soc., No. 1–2, 1955, pp. 137–138.
14. **A. M. Sakharov and I. P. Skibida**, Oxidation Commun., Vol. 1, No. 1, 1979, pp. 113–120.
15. **O. S. Fedorova and V. M. Berdnikov**, Izvestia AN SSSR, ser. khim., No. 4, 1979, pp. 745–749.
16. **E. S. Rudakov**, "Reactsii alkanov s okisliteliami, metallokompleksami i radicalami v rastvorakh." (The Reaction of Alkanes with oxidants, Metallcomplexes and Radicals in Solutions) Kiev, "Naukova Dumka," 1985, p. 88.

OXIDATION OF ORGANIC AMIDE IONS BY MOLECULAR OXYGEN

S. S. Shashin,* O. N. Emanuel, I. P. Skibida
Institute of Chemical Physics,
the Russia Academy of Sciences, Moscow

Abstract—Amide ions generated from secondary aromatic amines with pK_a>21 are readily oxidized by molecular oxygen at room temperature in a system KOH + 18-crown-6 + DMSO. On the basis of kinetic data it has been established that the reaction rate measured by oxygen consumption is determined by two main factors: the extent of amine ionization and the reactivity of the amide ions towards molecular oxygen, which change in opposite directions with increasing pK_a of the amines. The major paramagnetic and diamagnetic products of alkaline oxidation of secondary aromatic amines have been identified, and the sequence of their formation has been studied. A mechanism for oxidation of aromatic amide ions by molecular oxygen is put forward and discussed.

1. INTRODUCTION

Ionization of organic compounds in strongly basic media is a quite efficient and simple way of their activation with respect to molecular oxygen. At present this technique is being more and more widely used for selective oxidation of organic compounds under mild conditions, in great measure owing to the development of the methods of phase transfer catalysis which allow even very weak acids (pK_a about 30–35) to be ionized in readily available strongly basic systems {alkali + phase transfer catalysts (crown esters and their analogs + aprotic solvents} [1,2].

However, the range of substrates for which the mechanism and the preparative capacity of anionic oxidation have been studied in detail is mainly restricted to the C-H acids, whereas the oxidation patterns for organic anions with the charge centered on heteroatoms including nitrogen (amide ions) have been investigated very scantily [3-5].

The major oxidation products of primary aromatic amines (substituted anilines and naphthylamines) in alkaline media are derivatives of azobenzene and phenothiazine, respectively, and also nitroso- and nitro compounds [3–5]. To explain the composition of the oxidation products of amide ions, it has been suggested that the primary reaction products are aminyl radicals which can participate in C-N or N-N recombinations.

Secondary aromatic amines with reactive C-H bonds in the α position with respect to the amino group are, in the presence of bases, oxidized with cleavage of the C-H and C-N bonds [5].

With the use of EPR, the alkaline oxidation products of diphenylamine and phenyl-2-naphthylamine have been shown [6] to contain paramagnetic particles including nitroxyl radicals.

Analysis of the literature reveals the lack of generally accepted notions on the mechanism of amide ions oxidation by molecular oxygen.

In the present work, the kinetic patterns of the oxidation of organic amines by molecular oxygen in strongly basic media have been studied to elucidate the mechanism of auto-oxidation of amide ions.

Generation of amide ions from most aromatic amines containing electron-withdrawing substituents presents no great difficulties; moreover, ionization of substituted diphenylamines and anilines can be conveniently used to determine the values of the acidity function H_- in strongly basic media [7]. However, the alkaline oxidation of amines with activated C-H bonds often proceeds not at the nitrogen but at the carbon atom [5]. Besides, the amide ions generated from primary amines readily enter into reactions of condensation and substitution with the intermediate products of their oxidation [3–5].

These circumstances impose certain limitations on the choice of organic amines whose oxidation in alkaline media could yield unequivocal information on the kinetics and the mechanism of amide ions interaction with molecular oxygen involving the nitrogen atom. Therefore in this work the oxidation in strongly basic media has been studied with secondary amines (SA) containing no activated C-H bonds or other groups capable of being ionized and oxidized concurrently with the amino group. Use was mainly made of the derivatives of diphenylamine, (2,2-dimehyl) tetrahydroquinoline, and (2,2,6,6-tetramethyl) piperidine.

2. Experimental

Diphenylamine and morpholine were commercial reagent-grade products, and SA I-XII (structures presented below) were synthesized in the laboratory and their purity was not less than 98%.

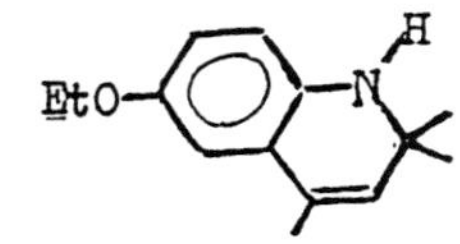

I

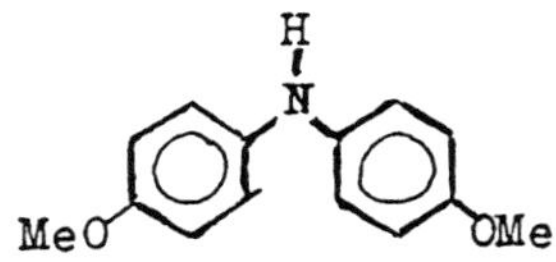

II

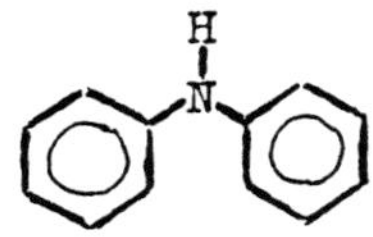

III

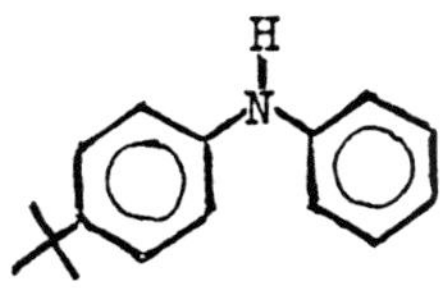

IV

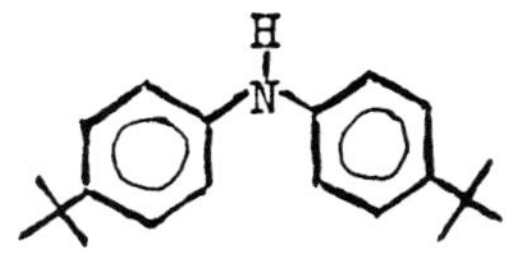

V

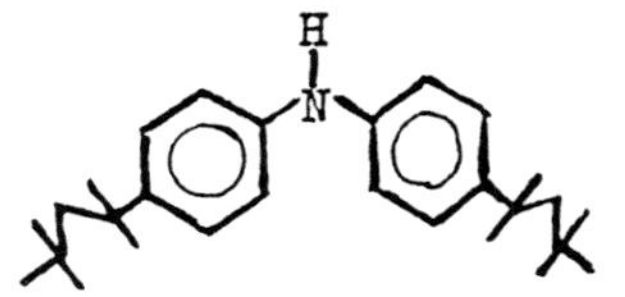

VI

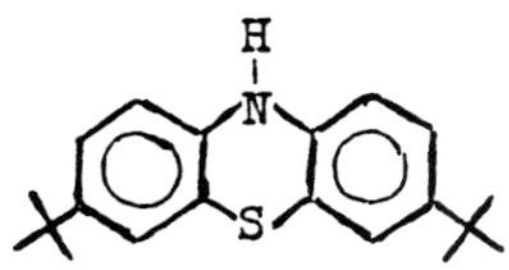

VII

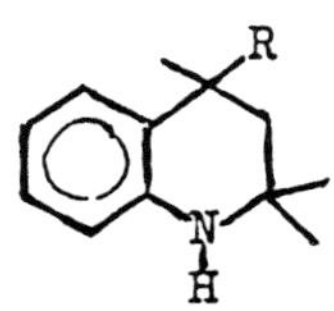

VIII

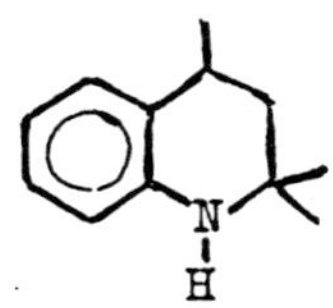

IX

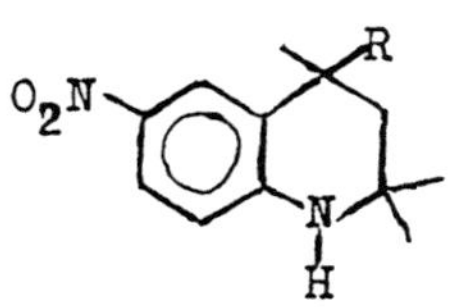

X

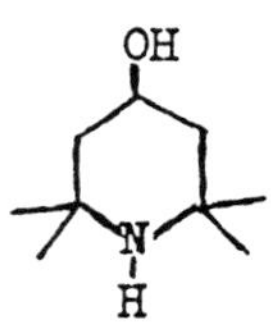

XI

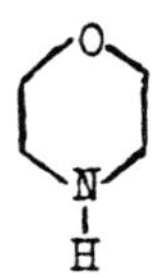

XII

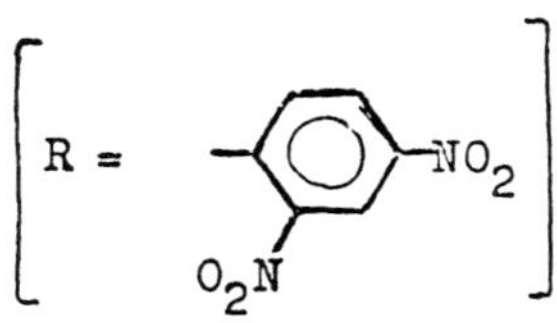

Aprotic solvents were purified by distillation or by passing through a column with chromatographic alumina, and stored over dessicants (anhydrous sodium sulfate or calcium chloride). Dimethyl sulfoxide was distilled under reduced pressure and stored over 4 Å molecular sieves. The water content in the purified solvent did not exceed 0.5%. KOH was homogenized in benzene essentially as described earlier [8].

The kinetics of O_2 consumption were studied in a volumetric unit at constant pressure. In a typical kinetic experiment, 0.2 mmole of a SA and 4.5 ml DMSO were introduced into a 25–100-ml cell equipped with a magnetic stirrer and, after purging the cell with oxygen or an argon-oxygen mixture, 0.5 ml of 1 M KOH in benzene was added. The partial oxygen pressure in the gas mixtures in the reaction course remained practically constant owing to a large volume excess of the gas in the cell; the changes in O_2 concentration throughout the reaction in all experiments did not exceed 5% of the initial value.

The absorption spectra of the compounds in the UV and visible regions were recorded in a Beckman UV 5270 spectrophotometer.

The extent of amine ionization in strongly basic media was determined in Ar according to the standard analytical procedures.

3. Paramagnetic Products of Secondary Amines Oxidation in Strong Basic Media

With most secondary aromatic amines studied, oxidation in strongly basic media is accompanied by formation of paramagnetic particles (Figure 1a–c, Table 1). The paramagnetic particles have been identified as nitroxyl radicals by comparing the EPR spectra observed with the data available in the literature on the hyperfine splitting (HFS) of the spectra of nitroxyl and aminyl radicals of known structure.

Collation of the EPR spectra of SA oxidates at various extents of reaction reveals no differences except broadening of the lines at increasing concentration of the nitroxyl radicals, which allows them to be regarded as the major paramagnetic products of the oxidation of secondary aromatic amines both at the early stages of the reaction and at deep conversion of the amines. The yield of nitroxyl radicals in alkaline SA oxidation is substantially dependent on their stability and reaction conditions, varying from fractions of a percent (amines I, III, IV, VII) to tens of percent (amines II, V, VI, IX).

Besides nitroxyl radicals, other nitrogen-containing paramagnetic particles can be observed under certain conditions in the course of alkaline SA oxidation. Thus when the reaction is interrupted at an early stage by discontinuing the oxygen supply (e.g., by purging the reaction mixture with argon), consumption of nitroxyl radicals is demonstrable by EPR. The rate of this process depends on the solvent and is especially high in DMSO. In some cases (amines IV–VI) the consumption of nitroxyl radicals is partly or completely reversible; when oxygen is blown through the reaction mixture there is an abrupt increase in the EPR signal intensity. A similar picture is observed in oxidation of amines I and II: discontinuation of the oxygen supply also entails the disappearance of the EPR signal from the nitroxyl radical; however, in these cases on the background of the nitroxyl signal there appear signals from new paramagnetic particles that have been identified as corresponding aminyl radicals by their HFS parameters and the characteristic ratio of the HFS isotropic (a^N) and anisotropic (b^N) constants (see Figure 1b, Table 2). According to the literature, for aromatic nitroxyl radicals the typical a^N/b^N values are 1.27 to 1.36 [9], whereas for aminyl radicals $a^N/b^N = 1.05 \pm 0.05$ [10]. The values $a^N/b^N = 1.33 \pm 0.05$ determined here for the radicals formed from compounds II–VII under aerobic conditions confirm that they have been justly referred to nitroxyls, while the value of this parameter for the new paramagnetic particle formed under anaerobic conditions is close to those found for aminyl radicals.

Table 1. Occurrence and EPR spectral parameters of nitroxyl radicals during alkaline oxidation of SA. DMSO, 25° C, Ar.

Amine	Radical detected (+/–)	HFS constants (gauss)		
		a^N	$a^H_{ortho}\left(=a^H_{para}\right)$	a^H_{meta}
I	+	12.0	2.7	
II	+	10.6	1.9	0.94
III	+	10.1	1.9	0.93
IV	+	10.1	1.9	0.95
V	+	10.2	1.9	0.95
VI	+	10.4		
VII	+	9.8	2.3	
VIII	—			
IX	+	12.1	2.8	
X	—			
XI	traces	14.8		
XII	—			

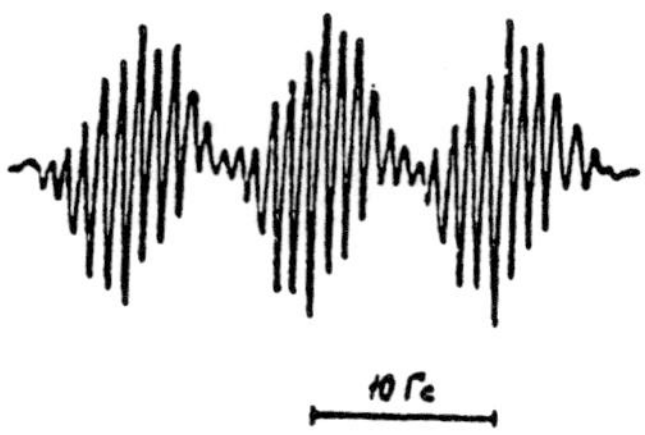

Figure 1a

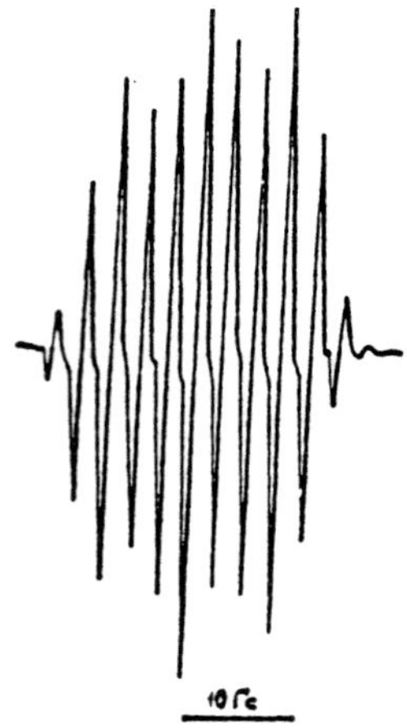

Figure 1b

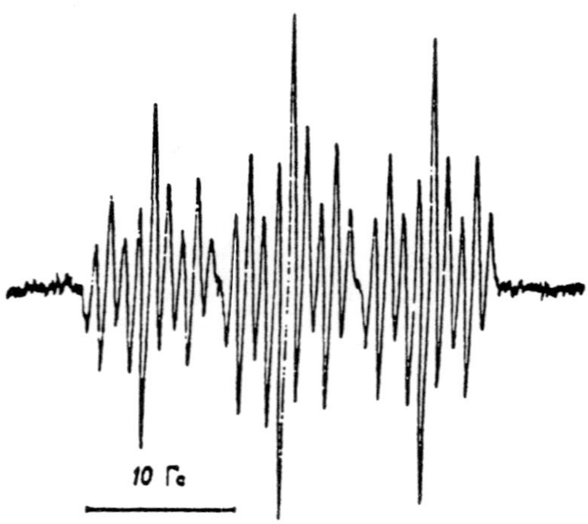

Figure 1c

Figure 1. EPR spectra of paramagnetic particles observed during SA alkaline oxidation (DMSO, 25° C, Ar).
(a) 4,4'-ditertbutyl diphenylamine-N-oxyl.
(b) 4,4'-dimethoxydiphenylaminyl radical;
(c) anion radical of 4-tertbutyl nitrobenzene.

Table 2. EPR spectral parameters of paramagnetic particles observed during SA oxidation (DMSO, 25° C, Ar).

Amine	HFS constants (gauss)				Radical identified
	a^N	a^H_{ortho} $\left(= a^H_{para}\right)$	a^H_{meta}	a^N/b^N	
I	12.0	2.7		1.33 ± 0.05	$NO^{\cdot}$
	7.0	2.7		—	$N^{\cdot}$(*)
II	10.6	1.9	0.94	1.33 ± 0.05	$NO^{\cdot}$
	7.8	2.6		0.87 ± 0.07	$N^{\cdot}$(*)
V	10.2	1.9	0.95	1.33 ± 0.05	$NO^{\cdot}$
	10.15	3.15	1.05	1.19 ± 0.07	$N^{\cdot}$(**)

(*) Observed upon flushing the oxidate with argon at early stages of the reaction.

(**) Observed when the reaction is conducted in the presence of $2 \cdot 10^{-4}$ mole cobalt stearate per liter.

Paramagnetic particles of a third type, identified as anion radicals of 4-nitro-tertbutylbenzene (Figure 1c), are observed in the course of alkaline oxidation of amine V in DMSO in the presence of catalytic amounts of cobalt stearate. The HFS parameters of the EPR spectrum of this particle are presented in Table 2. It is interesting to note that the value $a^N/b^N = 1.2$ appears to be typical enough of the anion radicals of aromatic nitro derivatives, because we have obtained the same value for the nitrobenzene anion radical as well.

4. Diamagnetic Products of Secondary Amines Oxidation in Strong Basic Media

The main data on the qualitative composition of the low-molecular-weight reaction products have been obtained by gas chromatography/mass spectrometry. The most detailed study has been made of the oxidation products of amine V, which are listed in Table 3. Analogous products are observed upon alkaline oxidation of amines II and IX, although in the latter case there are no C-N cleavage products (see Table 3). The oxidate of amine IX in DMSO displays absorption

Table 3. Products of alkaline oxidation of SA in DMSO identified by GS/MS

Amine	Major ions m/z (intensity, %)	Compound identified	(mol. mass)
V	149 (25), 134 (100), 118 (13), 106 (22), 91 (12), 78 (12)		(149)
	179 (12), 164 (100), 136 (11), 134 (13), 115 (24), 91 (30), 78(31), 62(30)		(179)
	295 (80), 280 (100), 264 (32), 249 (45), 134 (28)	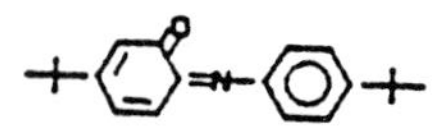	(295)
	311(75), 296(100), 280(15), 266 (12), 141 (25)	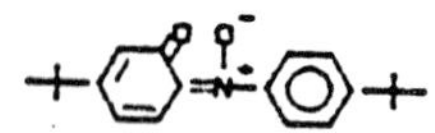	(311)
	150 (50), 136 (98), 119 (14), 95 (36), 91 (68), 77 (100)	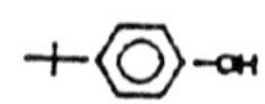	(150)
II	123, 107, 80, 65,62	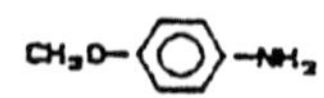	(123)
	243, 228, 213, 198, 81	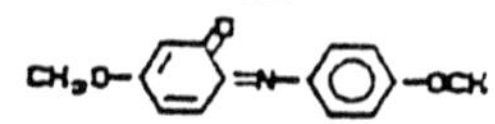	(243)
	259, 244, 229, 228, 213	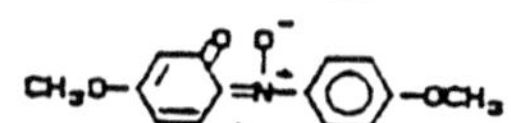	(259)
IX	189, 174, 160, 158, 144, 132, 130	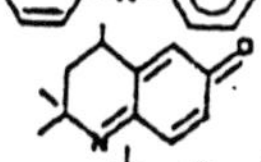	(189)
	205, 190, 175, 174, 60	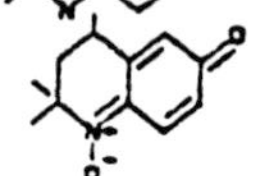	(205)

spectra with maxima at 295, 375, and 520 nm. The 295-nm and 375-nm absorption bands are in accord with the absorption spectra of quinone imines (λ_{max} = 290 nm [11] and quinone nitrons (λ_{max} = 371 nm [12]) also identified by GC/MS. The broad weak band around λ_{max} = 520 nm can supposedly be due to the absorption of the oligomeric products of amine N-N and C-N condensation.

A characteristic feature of alkaline oxidation of diarylamines in DMSO is the intense green color of solution, which develops in a few minutes after the onset of reaction and either persists throughout the oxidation (amine II) or for some time (amines III, IV), or quickly turns to red-brown (amines V, VI). In all cases the green color is associated with formation of oxidation products having a broad absorption band with a maximum about 640 nm, which can be tentatively attributed to the diarylamine C-N and N-N condensation products.

Thus, alkaline oxidation of secondary amines yields a broad set of products and intermediate particles; kinetic studies have been undertaken to elucidate the pathways of their formation and their role in the oxidation process.

5. Kinetics of Oxidation of Secondary Aromatic Amines in the Presence of Bases

The main patterns of SA oxidation by molecular oxygen have been studied in a strongly basic system KOH + 18-crown-6 + DMSO. The oxidation rate of tested amines, determined from the kinetics of oxygen consumption at room temperature, decreases in the order

$$I>II>III>IV>V,VI>VII>VIII>IX>X>>XI,XII$$

Amines XI and XII are practically not oxidized under such conditions, apparently because their NH-acidity is extremely low and the basicity of the system is not sufficient for effective ionization. Secondary aromatic amines I–IX are rapidly oxidized in the presence of strong bases already at room temperature. The series given above shows that the SA oxidation rate is substantially dependent on the nature of the substituent in the aromatic ring, being enhanced by strong electron-donor substituents (alkoxy groups) and attenuated by electron-withdrawing ones (nitro group).

6. Concentration of Amide Ions

To determine the concentration of amide ions in solution, the SA ionization equilibrium in KOH + 18-crown-6 + DMSO has been assessed spectrometrically. Some characteristics of the absorption spectra of the SA and respective amide ions as well as the NH-acidity values for a number of SA are given in Table 4.

The equilibrium ionization of secondary amines (AmH) in DMSO

$$AmH + OH^- \; Am^- + H_2O$$

is described by the equation

$$K_{eq} = [Am^-]_{eq}/[AmH]_{eq} \cdot [OH^-]_{eq} \quad (1)$$

Water formation upon SA ionization does not appreciably affect the K_{eq} because of the substantial water content in the solvent (about 0.3 mole/1) which remains practically constant, since ionization produces only 10^{-3} to 10^{-2} mole of water per liter. Yet, an equilibrium constant determined for one concentration of the alkaline additive does not, as a rule, provide adequate agreement between the calculated and the experimental concentrations of amide ions at other KOH concentrations. This appears to be so because the basicity of the medium is strongly influenced by the additions of water and methanol which are inevitably present in the alkaline additive (KOH + 18-crown-6) prepared according to [8].

Table 4. Absorption maxima and molar coefficients of SA and corresponding amide ions, and NH-acidity of SA in DMSO.

Amine (AmH)	l_{max}^{AmH}, nm	ε^{AmH}	$l_{max}^{Am^-}$, nm	ε^{Am^-}	pK_a
II	290	21,200	363	29,000	23.8
III	289	22,500	373	28,600	22.4
IV	290	23,600	370	29,000	22.8
V	291	27,500	371	30,200	23.3
VII	320	6,500	385	8,400	22.1
X	364	20,000	427	22,200	21.2

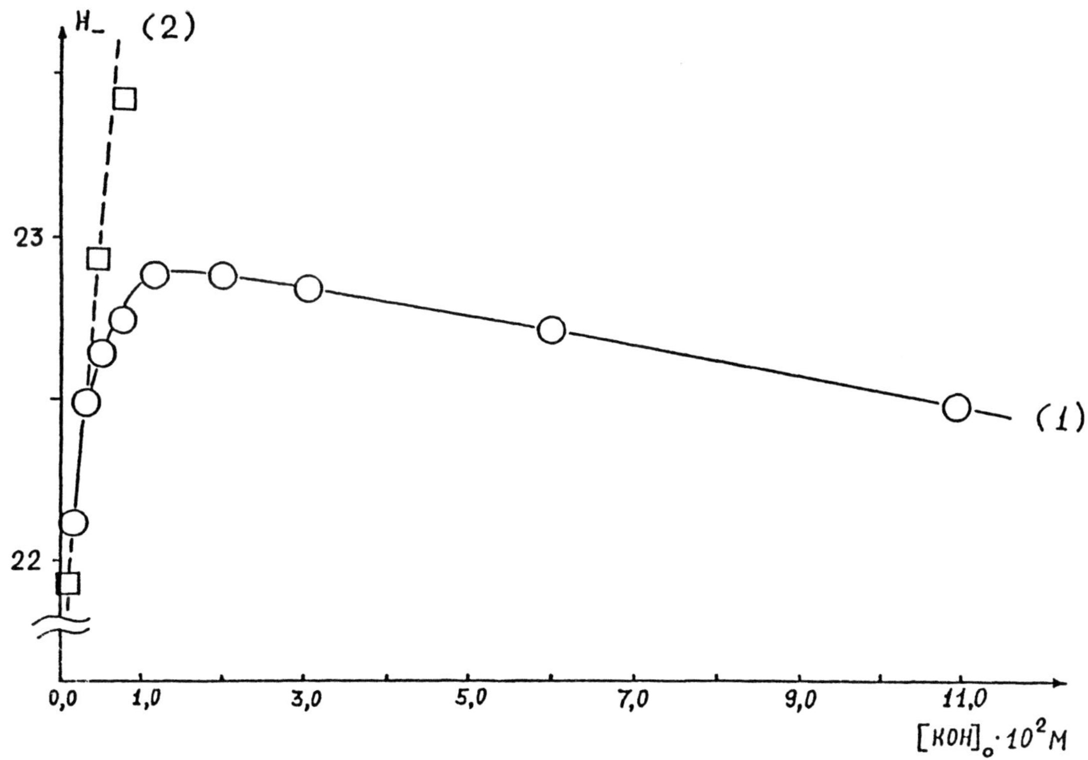

Figure 2 Dependence of the basicity of systems (1) DMSO + KOH + 18 crown 6 and (2) DMSO +$Me_4N^+OH^-$ on the amount of base added.

Such a nonlinear pattern of the changes in the basicity of the medium with increasing KOH concentration is illustrated in Figure 2, where the basicity of the medium is expressed as the values of the acidity function H_- determined from the spectrophotometric data on the ionization of indicator amounts of amines with known pK_a:

$$H_- = pK_a + \lg ([Am^-]_{eq}/[AmH]_{eq})$$

Here the indicator was diphenylamine with $pK_a = 22.44$ [7]. Figure 2 shows that the basicity of the medium increases linearly with alkali concentration only at $[KOH]_0 < 5 \cdot 10^{-3}$ M. At higher alkali concentrations the linearity is lost, at $[KOH]_0 = 1.5 \cdot 10^{-2}$ M the function reaches a maximal value of 22.9 and then appreciably decrease. This pattern is in good qualitative agreement with the changes in the relative amounts of water in DMSO ($[H_2O]_{DMSO}$) and of OH-acids (water and methanol, designated XOH introduced with the alkaline additive). Indeed, at $[KOH]_0 < 5 \cdot 10^{-3}$ M, when $[XOH] << [H_2O]_{DMSO}$, the solvent composition practically does not change upon addition of KOH, and the basicity of the medium increases with the amount of the alkaline additive. At $[KOH]_0 > 2 \cdot 10^{-2}$M, when [XOH] and $[H_2O]_{DMSO}$ are comparable, the solvent composition is significantly altered by the alkaline additive, and the basicity of the medium declines despite the increasing alkali concentration.

For comparison, Figure 2 also displays the changes in the basicity of the medium where the base is tetramethyl-ammonium hydroxide pentahydrate (Janssen Chimica) which does not appreciably change the solvent composition up to $[OH^-]$ of $(1–2) \cdot 10^{-2}$ M, and the basicity increases linearly with the amount of the base added.

Thus, in calculating the concentration of amide ions account should be taken of the water content in the solution. Experimentally, $[Am^-]$ can be most conveniently determined with the use of the dependence of H_- on [KOH], taking account of the relation (2) between $[Am^-]$ and H_- which holds well on condition that $H_- << pK_a$ or $[AmH]_0 << [KOH]_0$.

7. Kinetics of Substituted Diphenylamines Oxidation

To study the kinetics of alkaline oxidation of SA, we chose 4,4'-ditertbutyl diphenylamine (V) as a model compound.

The activation energy for alkaline oxidation of V in DMSO is 6.5 kcal/mole. The oxidation rate increases linearly with partial oxygen pressure in the range 0.2 to 1.0 atmosphere, testifying to the involvement of O_2 in a rate-limiting step of the process (see Figure 3, dependence 1).

The reaction displays fractional order in respect of amine and base concentrations: approximately 0.3 in the base concentration and 0.4 in the amine concentration. The reaction orders obtained are in accord with a suggestion that the rate-limiting step is the interaction of oxygen with the amide ion formed in the equilibrium step of amine ionization.

Thus, the rate of 4,4'-ditertbutyl diphenylamine oxidation in DMSO + KOH + 18-crown-6 is described by the equation

$$W^{o}_{O_2} = k\left[Am^-\right]_o \cdot P_{O_2} \tag{3}$$

with an average k = (5.5 ± 0.3) · 10^{-3} sec^{-1} atm^{-1} (25° C).

The nature of the solvent strongly affects the SA oxidation rate. In proton-donor solvents (water, primary alcohols) at room temperature, SA are extremely slowly oxidized even at [KOH] about 1 mole/liter.

In aprotic media the oxidation rate of V increases with solvent polarity in the order (dielectric constants given in parentheses): dioxane (2.2) < dimethoxyethane (7.2) << tetrahydrofurane (7.8) < benzene (2.28) < toluene (2.34) < chlorobenzene (5.62) < acetonitrile (37.5) < DMSO (46.7). The correlation between the oxidation rate and the solvent polarity is strictly observed only among solvents of close chemical structure (ethers, or aromatic hydrocarbons); that the aromatic solvents "fall out" of the general order is probably due to specific solvatation of amide ions by the p-donor molecules enhancing their reactivity.

In accordance with equation (3), the initial rate of SA oxidation is determined by two factors: the concentration of amide ions, and their reactivity towards molecular oxygen (k). Having estimated the concentration of amide ions generated from SA at certain basicity of the medium, and using the initial rates of oxygen consumption to compute the apparent rate constants of the reaction, $k_{app} = k \cdot P_{O_2}$, one can compare the reactivity of amide ions of different structure towards molecular oxygen (see Table 5).

For amines I and IX the pK_a values cannot be determined by the method used in this work because of their low NH-acidity. However, with an accuracy sufficient for our purposes the pK_a values for these amines can be estimated from the data available in the literature on the NH-acidity of related compounds and on the influence on pK_a of the substituents at the nitrogen atom and at the phenyl rings of aromatic amines. Thus taking the NH-acidity value of amine X and considering that introduction of a para-nitro group into the phenyl

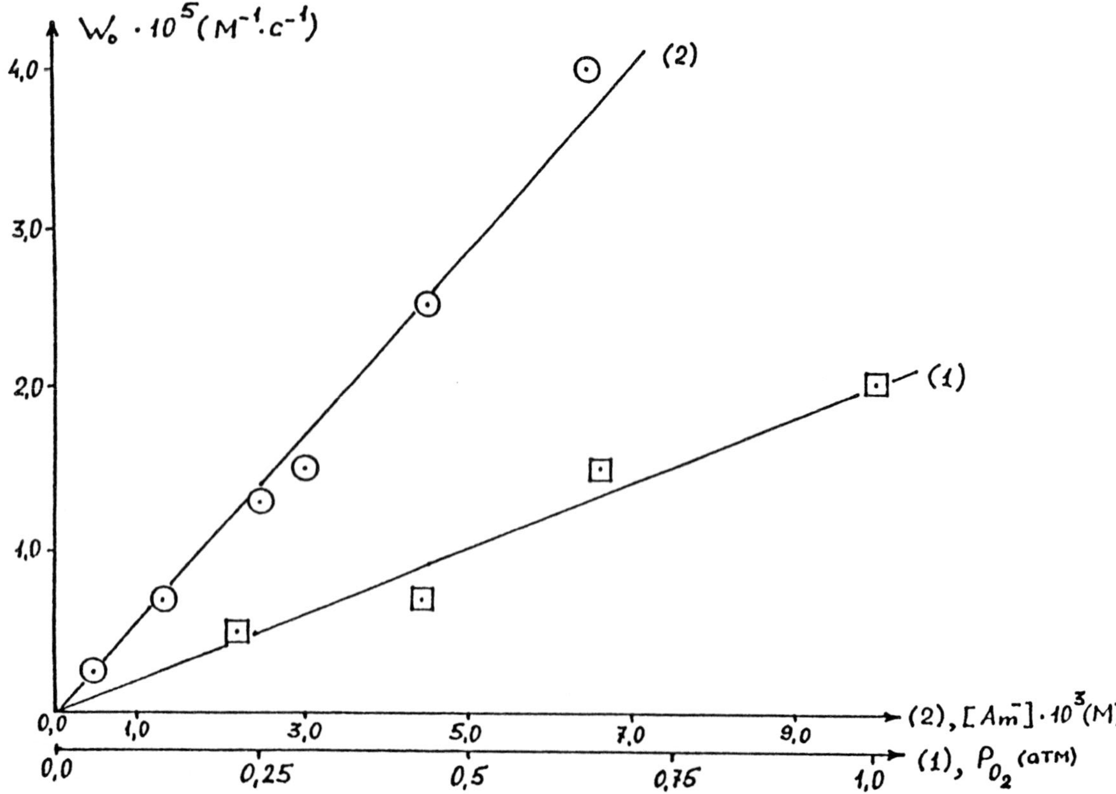

Figure 3 Dependence of the inital rate of alkaline oxidation of 4,4'-diterbutyl diphenylamine on (1) partial oxygen pressure and (2) amide ion concentration. DMSO, 25° C; $[KOH]_0$ = 0.023 M; (1) $[AmH]_0$ = 0.02 M; (2) $[AmH]_0$,<< $[KOH]_0$

Table 5. Kinetic parameters of SA oxidation in KOH + 18-crown-6 + DMSO. $[SA\}_o = 0.04$ M, $[KOH]_o = 0.1$ M, $P_{O_2} = 1.0$ atm, $t^o = 25°$ C.

Amine	pKa	$[Am^-]_o$, M (*)	$W_o \times 10^5$, $M^{-1}\,sec^{-1}$	k_{app}^{AmH}, $sec^{-1}\,atm^{-1}$
I	29–30	$(1.3–13) \cdot 10^{-9}$	18.0	$(1.4–14) \cdot 10^4$
II	23.8	$2.6 \cdot 10^{-3}$	17.0	$6.5 \cdot 10^{-2}$
III	22.4	$2.3 \cdot 10^{-2}$	6.3	$2.7 \cdot 10^{-3}$
IV	22.8	$1.4 \cdot 10^{-2}$	4.7	$3.4 \cdot 10^{-3}$
V	23.3	$6.5 \cdot 10^{-2}$	4.0	$(5.5 \pm 0.3) \cdot 10^{-3}$ (**)
VII	22.1	$2.9 \cdot 10^{-2}$	3.9	$1.3 \cdot 10^{-3}$
IX	28–29	$(1.3–13) \cdot 10^{-8}$	3.3	$(2.5–25) \cdot 10^2$
X	21.2	$3.8 \cdot 10^{-2}$	0.5	$1.3 \cdot 10^{-4}$

(*) Values calculated according to equation (2).
(**) Average value obtained over a broad range of amine and alkali concentrations.

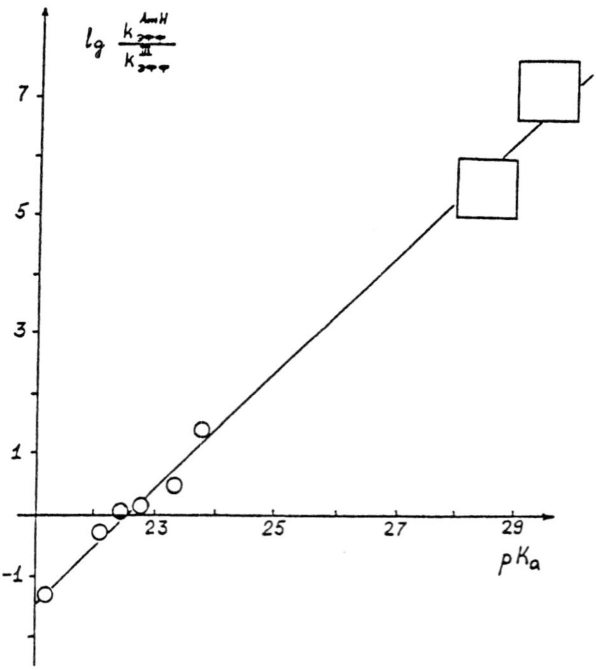

Figure 4. Dependence of the reactivity of amide ions towards molecular oxygen on the NH-acidity of corresponding amines.

substitutents of aromatic amins lowers the pK_a by 7–8 units; introduction of an alkoxy group raises it by 0.5–1.0; and replacement of a hydrogen of a primary amino group with an iso-alkyl substituent lowers the pK_a by 0.5–1.0 (see, e.g., [7], one can predict with reasonable certainty the ranges for the pK_a values of amines I and IX (see Table 5) and, consequently, the concentrations of the respective amide ions in media of known basicity.

As evident from Table 5, there is no positive correspondence between W_{O_2} either the extent of amine ionization or the reactivity of amide ions towards oxygen. The rate of alkaline oxidation of secondary aromatic amines is determined by a superposition of both factors. At the same time, there is a clear-cut correlation between the NH-reactivity of the amide ions towards O_2 and the decreasing NH-acidity (increasing pK_a) of the respective amines (Figure 4).

Analogous correlations are also quite well-known for many reactions involving carbanions, but are difficult to link up to a particular mechanism of the reactions under study, since such correlations have been observed both for nucleophilic substitution and addition as well as for single-electron oxidation of carbanions [13,14].

8. Kinetics of Formation and Consumption of Nitroxyl Radicals

During alkaline oxidation of secondary amines, EPR demonstrates either a continuous growth of or an extremal function of changing in the concentration of nitroxyl radicals formed in the process. The first pattern of kinetic behavior is typical for amines II, V, and VI producing stable nitroxyl radicals on oxidation, and the second pattern for amines I, III, IV, VII, and IX whose radicals are less stable.

We have used oxidation of amine V as a model reaction for studying the kinetics of the formation of nitroxyl radicals. When V is oxidized in DMSO at room temperature, the rate of nitroxyl radical accumulation increases with the initial base and amine concentrations and oxygen pressure. Nitroxyl radical formation exhibits the following reaction orders in respect of the components of the system $N_{O_2} = 0.9, n_{AmH} = 1.2, n_{KOH} = 0.7$.

The yield of the nitroxyl radical is greatly affected by the solvent. In the series of solvents tested (except DMSO) the nitroxyl accumulation rate and yield change in the same general order as the amine oxidation rate (the nitroxyl yield in 24 h is given in parentheses): chlorobenzene (65–70%) > toluene (45–50%) > tetrahydrofurane (30–35%). In DMSO the nitroxyl yield is only 10–12% although the amine oxidation rate as measured by oxygen consumption is maximal.

The reaction stoichiometry (the ratio of the amounts of oxygen consumed to nitroxyl radical formed) varies from 1 in chlorobenzene (at amine conversion less than 50%) to 10 in DMSO. Noteworthy is the association between a high radical yield and a low stoichiometry of oxygen consumption, which gives grounds for believing that there exists a reaction pathway whereby consumption of 1 mole O_2 per 1 mole of amine produces 1 mole of nitroxyl radical.

As indicated above, if SA oxidation is interrupted at an early stage by purging the reaction mixture with argon, disappearance of the accumulated nitroxyl radicals can be observed by EPR. We have suggested that electron transfer takes place under such conditions:

$$R_2NO^{\cdot} + R_2N^- \rightarrow R_2NO^- + R_2N^{\cdot}$$

The aminyl radicals thus formed usually perish at a high rate, but if they are actively generated and relatively stable they can be directly detectable by EPR (for instance, during oxidation of I and II as mentioned above).

This suggestion has been tested in a model system: amine V and the corresponding nitroxyl were dissolved in DMSO at a preset ratio in an atmosphere of inert gas, and the reaction was initiated by adding the alkali. The R_2N^- concentration was monitored spectrophotometrically, the $R_2NO^{\cdot}$ concentration by EPR. Under such conditions, the disappearance of $R_2NO^{\cdot}$ is described by the equation

$$W = k \cdot [R_2NO^{\cdot}] \cdot [R_2N^-]$$

with the rate constant $k = 7.5 \pm 0.4\ M^{-1} \cdot sec^{-1}$.

When V is replaced with II, under similar conditions EPR registers signals of aminyl radicals generated from II. The assumption that interaction between $R_2NO^{\cdot}$ and R_2N^- entails electron transfer from the amide ion to the radical is favored by the fact that $R_2NO^{\cdot}$ are regenerated when oxygen is supplied into the system. Such regeneration proceeds at a rate two orders of magnitude higher than that of radical accumulation during SA alkaline oxidation, and appears to take place at the expense of rapid oxidation of diarylhydroxylamine anions by oxygen:

$$R_2NO^- + O_2 \rightarrow R_2NO^{\cdot} + O_2$$

$$R_2NO^- + O_2 \rightarrow R_2NO^{\cdot} + O_2^{2-}$$

Provided a large excess of the amine to maintain a practically constant concentration of the amide ions, the cycle "$R_2NO^{\cdot}$ consumption under inert gas - $R_2NO^{\cdot}$ regeneration after an oxygen pulse" can be repeated many times (see Figure 5).

Thus, a high amide-ion concentration may attenuate the yield of stable nitroxyl radicals, and this is the probable cause of the abnormally low yield of nitroxyl radicals during oxidation of V in DMSO as compared with less polar solvents (less basic media).

The relatively stable aminyl radicals generated from amines I and II apparently not only recombine but also take part in reaction of cross-disproportionation with the nitroxyl radicals. This is evidenced by the fact that an EPR signal from the aminyl radicals evolves only after a virtually complete disappearance of the EPR signal from the nitroxyl ones, and corresponds to not more than a few percent of the initial nitroxyl concentration. The EPR signals of the aminyl radicals generated from amines I and II are observable in solution for several hours, and persist when oxygen is introduced into the system (in the case of amine II) or vanish without formation of an EPR signal of nitroxyl radicals (in the case of amine I).

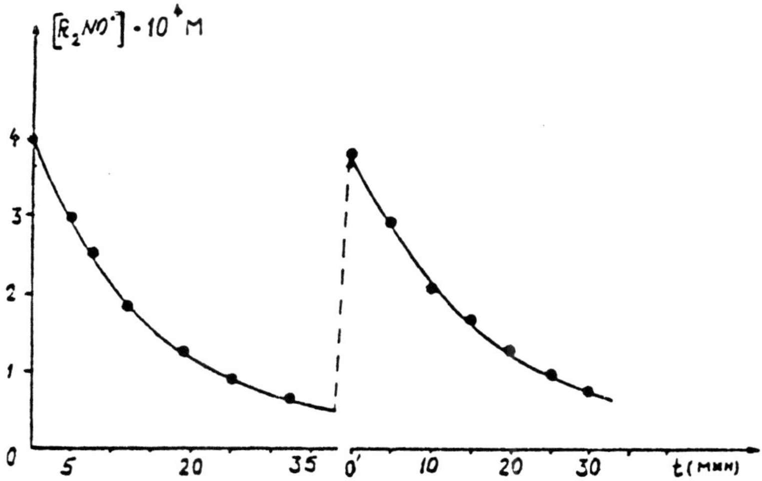

Figure 5. Changes in the nitroxyl radical concentration in the course of oxidation of amine V in DMSO + KOH + 18-crown-6 upon consecutive flushing of the oxidate with argon and oxygen. DMSO, 25° C, $[AmH]_0 = 8 \cdot 10^{-4}$ M, $[KOH]_0 = 1.6 \cdot 10^{-2}$ M. The broken line denotes the oxygen flush.

The presence of quinone imines and quinone nitrons among the oxidation products of II makes one suppose that with amine II (in contrast to amine V) the following sequence of reactions takes place under the conditions described above:

$$R_2N^{\cdot} + R_2N^{-} \rightarrow R_2NO^{-} + R_2NO^{\cdot}$$

$$2R_2N^{\cdot} \xrightarrow{\leftarrow} R_4N_2$$

$$R_2NO^{\cdot} \rightarrow R_2NH + \text{quinone imine}$$

Disproportionation of aminyl and nitroxyl radicals may proceed via a mechanism analogous to that known for the disproportionation of aromatic nitroxyl radicals [15] involving formation of quinone nitrons:

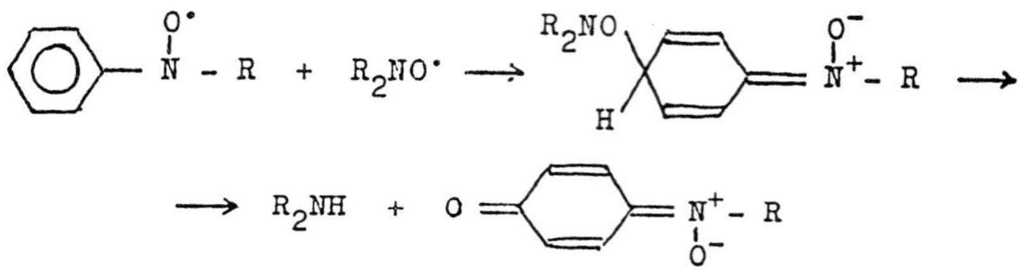

and may be the main source of quinone imines in the course of oxidation of secondary aromatic amines:

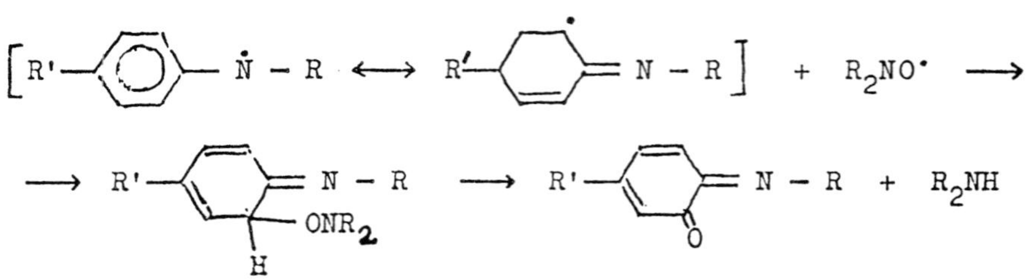

Although nitroxyl radicals are not registered in $R_2N^\cdot$-containing solutions even upon saturation with oxygen, the question of the possible formation of $R_2NO^\cdot$ by direct interaction of $R_2N^\cdot$ and O_2 remains open, in view of the possible rapid elimination of the newly formed nitroxyl radicals in accordance with the scheme above.

9. THE MECHANISM OF SECONDARY AROMATIC AMINE OXIDATION IN STRONG BASIC MEDIA

On the strength of the experimental data presented above on the set of products and the kinetics of alkaline oxidation of secondary aromatic amines containing no C-H bonds in the α -position with respect to the amino group, the following reaction mechanism can be proposed:

1. $R_2NH + OH^- \leftrightarrow R_2N^- + H_2O$
2. $R_2NH + OH^- \rightarrow RNH_2 + RO^-$
3. $R_2N^- + O_2 \rightarrow R_2NO_2^-$
4. $R_2NH + R_2NO_2^- \rightarrow R_2NOH + R_2NO^-$
5. $R_2NO_2^- \rightarrow RNO_2 + R^-$
6. $2R^- + O_2 \rightarrow 2RO^-$
7. $R_2NO^- + O_2 \rightarrow R_2NO^{\cdot} + O_2$
8. $R_2NOH + O_2 \rightarrow R_2NO^{\cdot} + HO_2^-$
9. $R_2NO^{\cdot} + R_2N^- \rightarrow R_2NO^- + R_2N^{\cdot}$
10. $R_2N^{\cdot} + R_2N^{\cdot} \rightarrow$ recombination products
11. $R_2N^{\cdot} + R_2NO^{\cdot} \rightarrow R_2NH +$ quinone imine
12. $R_2NO^{\cdot} + R_2NO^{\cdot} \rightarrow R_2NH$
    $+$ quinone nitron.

At the first step of the reaction, the amine is ionized to form an amide ion which undergoes subsequent oxidation. If the amine is ionized slowly (e.g., during oxidation of amines with a $pK_a > 25$, or when the reaction is conducted in weakly polar aprotic solvents) this step may limit the oxidation rate.

In most cases the rate-limiting step in the oxidation of secondary aromatic amines is the interaction of the amide ion with oxygen (step 3). The data in the literature on the mechanisms of interaction between the anionic forms of substrates with O_2 give grounds for suggesting two possible routes for this step.

(3a)

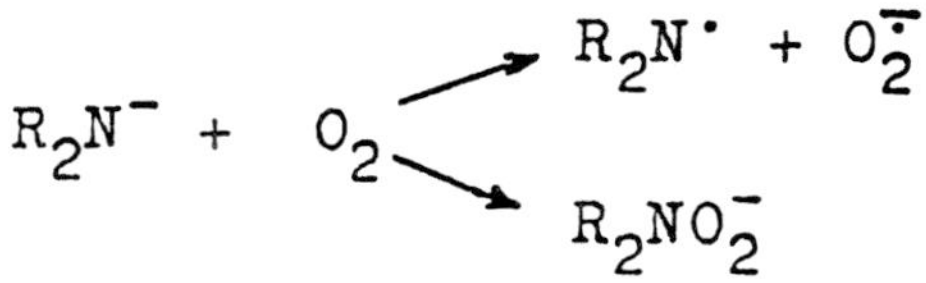

(3b)

It is believed that during oxidation of C-H acids, alcohols, ketones, or sulfides in alkaline media the key step of the reaction is the interaction of carbanions (or sulfide ions) with O_2 proceeding via a single-electron mechanism with formation of free radicals (route (3a); the reaction is markedly accelerated in the presence of catalytic amounts of strong electron acceptors [16].

With secondary aromatic amines, special experiments have shown that introduction of such electron acceptors as $PhNO_2$, $R_2NO^{\cdot}$, Co(3+), Cu(2+), or Fe(3+) practically does not affect the oxidation rate. This prompts a suggestion that during oxidation of amide ions, unlike carbanions, the interaction with oxygen takes place via the nonradical route (3b). This suggestion is also favored by the following arguments:

- no aminyl radicals are detectable by EPR during oxidation;
- aminyl radicals are apparently not precursors to nitroxyl radicals, since interaction of anaerobically generated $R_2N^{\cdot}$ radicals with O_2 fails to produce $R_2NO^{\cdot}$ even in those cases when oxidation of R_2NH in the presence of bases entails accumulation of large amounts of nitroxyl radicals in the system.

Two isomeric forms can be envisaged for the intermediate anion

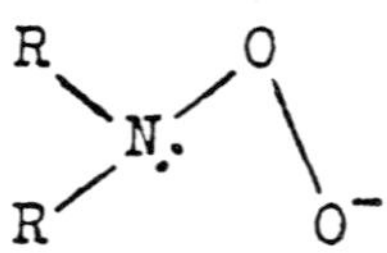

and

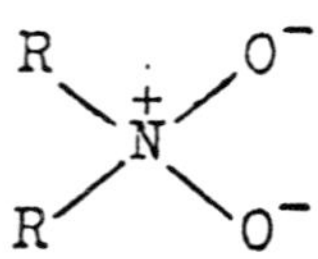

but at present it is impossible to decide between them. The nitroxyl radicals may be formed in the sequence of reactions (4,7,8) the first of which involves the anion $R_2NO_2^-$ (in one of the possible forms) as the donor of the oxygen atom.

The aminyl radicals are most probably secondary products of the reaction formed in step (9), which has been proved above to be possible.

As can be seen in Table 3, the oxidates contain products of the C-N bond cleavage (substituted anilines, nitrobenzenes, and phenols) which we believe to be formed according to overall reactions (2,5,6). Reaction (2) is appreciable only in the system DMSO + KOH + 18-crown-6, especially at elevated temperatures, and does not involve an oxidant since it takes place both under oxygen and under inert gas. The phenolate anions formed thereby are relatively stable to alkaline oxidation and do not affect the SA oxidation rate.

Splitting of the intermediate particles $R_2NO_2^-$ (step 5) may be the source of substituted nitrobenzenes found in small amounts among the diarylamine oxidation products. This reaction can probably be catalyzed by transition metal compounds, since in their presence the yield of nitro compounds (registered as the corresponding anion radicals) is substantially enhanced while the yield of nitoxyl radicals is attenuated.

The carbanions formed in step (5) are in all probability rapidly oxidized to substituted phenols according to the overall reaction (6).

The scheme proposed here provides a qualitatively adequate description of the set of products and the kinetic patterns of SA oxidation in strongly basic media.

10. REFERENCES

1. **M. Hiraoka**, Crown Compounds. Their Characteristics and Applications, Tokyo: Kodansha Ltd., 1982.
2. **R. Neumann and Y. Sasson**, J. Org. Chem., Vol. 49, No. 7, 1984, pp. 1282–1284.
3. **L. Horner and J. Dehnest**, Chem. Ber., Vol. 96, No. 3, 1963, pp. 786–797.
4. **E. Balogh-Hergovich, G. Speier, and E. Winkelmann**, Tetr. Lett., No. 37, 1979, pp. 3541–3542.
5. **A. A. Frimer, G. Aljadeff, and J. Ziv.**, J. Org. Chem., Vol. 48, No. 10, 1983, pp. 1700–1705.
6. **R. Hoskins**, J. Chem. Phys., Vol. 25, No. 4, 1956, pp. 788–789.
7. **D. Dolman and R. Stewart**, Can. J. Chem., Vol. 45, 1967, pp. 911–924.
8. **S. S. Shashin, O. N. Emanuel, and I. P. Skibida**, Bull. Acad. Sci. USSR D. Chem. Sci., Vol. 32, No. 10, 1984, pp. 2002–2006.
9. **A. M. Vasserman and A. L. Buchachenko**, Zhurnal Strukturnoi Khimini, Vol. 7, No. 5, 1966, pp. 673–679 (in Russian).
10. **A. L. Buchachenko and A. M. Vasserman**, Stabilnye radikaly (Stable Radicals), Moscow: Nauka, 1973, 408 pp. (in Russian).

11. **O. T. Kasaikina, T. V. Lobanova, D. V. Fentsov, and Yu. A. Ivanov**, Bull. Acad. Sci. USSR D. Chem. Sci., Vol. 32, No. 10, 1984, pp. 1994–1998.
12. **O. T. Kasaikina, T. V. Lobanova, and D. V. Fentsov**, Bull. Acad. Sci. USSR D. Chem. Sci., Vol. 32, No. 10, 1984, pp. 1998–2002.
13. **F. G. Bordwell and A. H. Clemens**, J. Org. Chem., Vol. 46, No. 5, 1981, pp. 1035–1037.
14. **A. A. Solovyanov and I. P. Beletskaya**, Uspekhi Khimii, Vol. 47, No. 5, 1978, pp. 819–846 (in Russian).
15. **A. R. Forrester and S. P. Hepburn**, J. Chem. Soc., Perkin I., No. 19, 1974, pp. 2208–2213.
16. **G. A. Russel, A.G. Bemis, E. J. Geels and E. G. Janzen**, Adv. Chem. Ser., Vol. 75, 1968, pp. 174–202.

Inhibition of Radical Chain Reactions by Complex Compounds of Copper(2+) and Nickel(2+) with Stable Nitroxyl Radicals

V. N. Vetchinkina
Institute of Chemical Physics,
The Russia Academy of Sciences,
Moscow, Russia

I. P. Skibida
Institute of Chemical Physics,
The Russia Academy of Sciences,
Moscow, Russia

S. V. Larionov and L. A. Petrina
Institute of Inorganic Chemistry,
Siberian Division of the Russia Academy of Sciences,
Novosibirsk, Russia

Abstract—Inhibition of oxidation and polymerization by stable nitroxyl radicals RNO of various structure and by their complexes with nickel(2+) and copper(2+) [$M(RNO)_2$] has been studied for oxidation of ethylbenzene and nonene–1 (70° C) and polymerization of styrene (60° C) in the presence of azobis(isobutyronitrile) as an initiator. In nonene–1 oxidation, the inhibitory activity of the nitroxyl radicals (RNO) is substantially enhanced when the anions are coordinated to copper or nickel atoms in $M(RNO)_2$ Inhibition takes place owing to the reaction of the $M(RNO)_2$ complexes with free alkyl radicals, an important part being played by "secondary" processes including the interaction of the inhibitor conversion products with peroxide radicals RO_2. Inhibition of styrene polymerization by the $M(RNO)_2$ complexes takes place owing to the interaction of free radicals with the RNO paramagnetic centers whose activity does not change upon coordination of the anions to copper or nickel.

1. INTRODUCTION

Stable nitroxyl radicals (RNO) are active acceptors of free radicals (R and RO_2) and in a number of cases can be used as inhibitors of oxidation and polymerization [1]. It has earlier been shown with various nitrogen-, sulfur-, and oxygen-containing compounds that incorporation of the inhibitors into complexes with metals can enhance their inhibitory activity [2–4].

2. RESULTS AND DISCUSSION

In the present work, the inhibition of oxidation and polymerization by stable nitroxyl radicals (I) and their complexes with nickel(2+) and copper(2+) (II) has been studied for oxidation the of ethylbenzene and nonene–1 and polymerization of styrene initiated with azo-bis(isobutyronitrile) (AIBN).

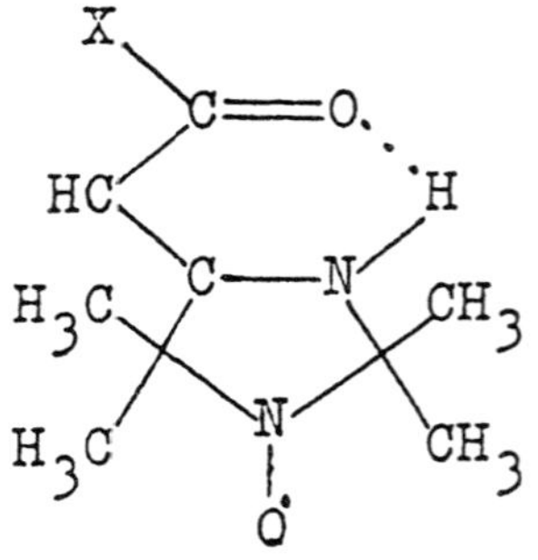

(I)

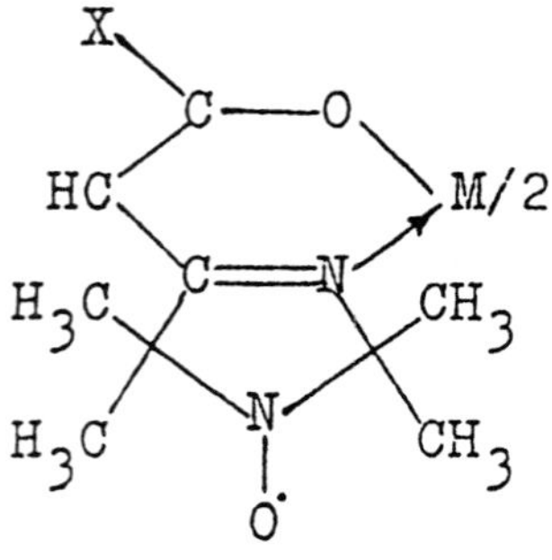

(II)

X = CH_3 (a), HC $\genfrac{}{}{0pt}{}{CH_3}{CH_3}$ (b), C_6H_5 (c); M = Ni(2+), Cu(2+).

Complexes II have been synthesized according to a published procedure [5].

Oxidation was carried out in a manometric unit, and the reaction rate was monitored by the oxygen consumption rate [6]. The polymerization rate was determined dilato–metrically [7]. The changes in the ligand (I) and complex $(II)_M$ concentrations in the course of oxidation were measured using EPR and UV spectroscopy.

Ligand (Ia) displays an absorption band at λ_{max} = 294 nm, whereas upon formation of complexes $(IIa)_{Ni}$ and $(IIa)_{Cu}$ this band is shifted towards longer wavelengths (λ_{max} = 305 nm) and a second absorption band appears at λ_{max} = 258 nm. The position of the absorption bands of the $(II)_M$ complex is independent of the nature of the metal.

Both the ligands and the $(II)_{Ni}$ and $(II)_{Cu}$ complexes display EPR triplets with the same *A* and *g* values, but the spectra of the $(II)_M$ complexes are much less intense and, according to [8], correspond to the forms in which there is no spin-spin interaction between the paramagnetic centers of the radicals; the portion of such complexes is rather small.

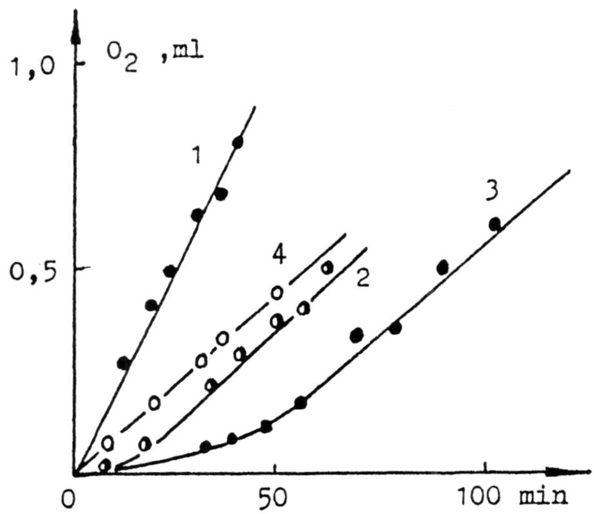

Figure 1. Kinetic curves of oxygen consumption during initiated oxidation of nonene-1 in (1,2) oxygen and (3,4) air: (1) control, (2,3) with $(IIa)_{Ni}$, and (4) with (Ia). $[AIBN]_0 = 5 \cdot 10^{-3}$ M, (Ia) = $(IIa)_{Ni} = 1.2 \cdot 10^{-3}$ M; temp. 70° C.

Oxidation of ethylbenzene proved to be only weakly retarded by addition of $(IIa)_{Ni}$ or $(IIa)_{Cu}$ at a concentration of $3.3 \cdot 10^{-4}$ M. The rate constants for the interaction of the inhibitors with free 1-phenylperoxide radicals have been calculated from the dependence of the oxidation rate on $(II)_M$ concentration according to [6] and in both cases have been found to be $k_7 = (1.4 \pm 0.1) \cdot 10^3\ M^{-1}\ sec^{-1}$.

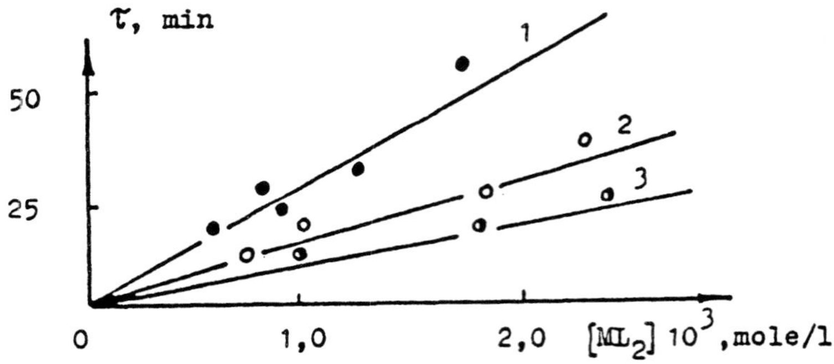

Figure 2. Dependence of the lag period in initiated nonene-1 oxidation on the concentration of $(IIa)_{Ni}$, curve 1; $(IIa)_{Cu}$, curve 2; and $(IIb)_{Ni}$, curve 3. $[AIBN]_0 = 5 \cdot 10^{-3}$M, temp. 70° C.

The $(II)_M$ compounds are much more inhibitory towards oxidation of nonene–1. As can be seen in Figure 1, introduction of $(IIa)_{Ni}$ into the reaction of initiated oxidation of nonene–1 results in the appearance of a distinct *lag* period τ in the kinetic curve of oxygen consumption (curve 2). Furthermore, not only the $(IIa)_{Ni}$ complexes but also their conversion products obviously have some inhibitory activity because after the *lag* period the rate of nonene–1 oxidation proves lower than that of initiated oxidation in the control (curve 1). The same qualitative pattern is also observed for nonene–1 oxidation in the presence of corresponding copper complexes.

A decrease of the O_2 content in the oxidizing gas (replacement of oxygen with air) prolongs the lag period in nonene–1 oxidation (curve 3).

The higher inhibitory activity of $(IIa)_{Ni}$ towards nonene-1 oxidation as compared with ethylbenzene oxidation and the increase in τ upon attenuation of the oxygen content in the oxidizing gas apparently indicate that $(IIa)_{Ni}$ is mainly capable of interacting with the R radicals rather than with the peroxide radicals RO_2.

The inhibitory effect of the stable nitroxyl radicals (Ia) is much less pronounced (curve 4 in Figure 1) than that of the corresponding complex compounds $(IIa)_M$. Addition of (Ia) only reduces the rate of nonene-1 oxidation in air but elicits no lag period in the kinetic curve of oxygen consumption, whereas in the presence of $(IIa)_{Ni}$ the lag in nonene–1 oxidation amounts to 33 min. The rates of nonene–1 oxidation in the presence of (Ia) and $(IIa)_{Ni}$ after the lag period are quite similar.

Table 1. Inhibition coefficients (f) in nonene–1 oxidation and polarographic oxidation potentials ($E_{1/2}$) of metal complexes.

	$(IIa)_{Ni}$	$(IIa)_{Cu}$	$(IIb)_{Ni}$	(Ia)
f	0.82	0.45	0.30	–
$E_{1/2}$	0.84	0.91	0.85	0.96

The efficiency of the inhibition of nonene-1 oxidation by the complex compounds $(II)_M$ depends on the metal but is also substantially altered by the introduction of substituents into the ligand.

Within the range of $(II)_M$ concentrations tested, τ is linearly dependent on the inhibitor concentration (Figure 2), making it possible to estimate the stoichiometric inhibition coefficient f (Table 1) within the framework of the classical concepts of the inhibition mechanism according to the equation $f = \tau \cdot W_i$ where W_i is the initiation rate and InH is the inhibitor concentration.

The compounds tested can be arranged in the order of increasing inhibitory activity: (Ia) < $(IIb)_{Ni}$ < $(IIa)_{Cu}$ < $(IIa)_{Ni}$.

In view of reports [9] that in quite a few cases the inhibitory activity of metal complexes correlates with their oxidation potentials, we have measured the $E_{1/2}$ values for the $(II)_M$ complexes as described in [9] using a Radelkis (Hungary) OH–102 polarograph. These data aregiven in Table 1, with the corresponding value for ligand (Ia) for comparison.

Collation of the f and $E_{1/2}$ values demonstrates that the correlation between the oxidation potential and the inhibitory activity only holds for similar complexes of different metals, such as $(IIa)_{Ni}$ *vs* $(IIa)_{Cu}$: expectably [9], a higher oxidation potential (i.e. lower oxidizability) results in a lower inhibitory activity of a metal complex. On the other hand, introduction of different substituents into the ligand abolishes the correlation; thus despite practically the same oxidation potential of the $(IIa)_{Ni}$ and $(IIb)_{Ni}$ complexes their antioxidative activity is markedly different (Table 1).

Since the $(II)_M$ conversion products also are inhibitors of oxidation (as evidenced by the lower oxidation rate after the lag period, see Figure 1), the observed lack of correlation between the oxidation potential and the inhibitory activity of metal complexes may be explained by the fact that the antioxidative activity of a complex compound is determined by the combined action of both the initial complex and its oxidation products, whereas the $E_{1/2}$ pertains only to the oxidizability of the initial compound. The substituents in the ligand mainly affect the inhibitory activity of the products of complex oxidation but have little effect on the $E_{1/2}$ value.

A study of the consumption of $(IIa)_{Cu}$ and $(IIa)_{Ni}$ reveals differences in their kinetic behavior during nonene-1 oxidation. When nonene–1 is oxidized in the presence of $(IIa)_{Cu}$, the intensity of the EPR spectra of the solution first declines and, having passed through a minimum, rises again (Figure 3a, curve 1). As the reaction comes out of the lag period, new paramagnetic particles $(IIa)^{ox}_{Cu}$ are formed in the system, and their spectrum is superimposed on the EPR spectrum of the initial solution. A similar pattern is observed for the kinetics of the optical density of the solution at 258 nm (Figure 3a, curve 2). At the wavelength corresponding to the ligand absorption (305 nm) the optical density of the solution decreases monotonously (curve 3).

When nonene-1 oxidation is inhibited by $(IIa)_{Ni}$ the kinetic pattern is totally different: in the early period the intensity of the EPR triplet signal of the solution sharply increases to a maximum and then declines (Figure 3b, curve 1). As it does so, the EPR spectrum remains unchanged in the course of oxidation, unlike the case with $(IIa)_{Cu}$. In the UV region, the optical density at 305 nm (absorption of the ligand) also passes through a maximum (Figure 3b, curve 3) while that at 258 nm (absorption of the complex) decreases) monotonously (Figure 3b, curve 2).

The observed similarity in the kinetic behavior of the system as followed with EPR (measuring only the particles that give a triplet signal) or UV spectroscopy (measuring the total $(II)_M$ content in the system) allows a conclusion that both methods produce a qualitatively correct picture. The analogy appears to be due either to the coincidence in the k_7 values for the two forms of the $(II)_M$ complex (with and without spin-spin interaction between the paramagnetic centers) or to an equilibrium between these forms.

On the other hand, the differences between $(IIa)_{Ni}$ and $(IIa)_{Cu}$ in the kinetics of the changes in EPR signal intensity and optical density of the solutions in the course of nonene–1 oxidation arenoteworthy, which is indicative of the different mechanisms of inhibition of nonene–1 oxidation by these complexes.

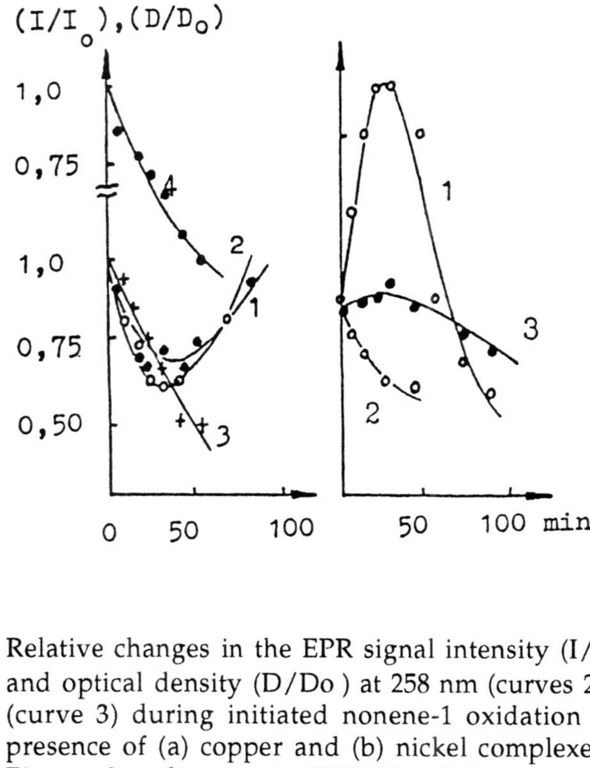

Figure 3. Relative changes in the EPR signal intensity (I/I_0) (curves 1) and optical density (D/Do) at 258 nm (curves 2) and 305 nm (curve 3) during initiated nonene-1 oxidation by air in the presence of (a) copper and (b) nickel complexes. Curve 4 in Figure *3a*: changes in EPR signal intensity in a mixture $(IIa)_{Cu}$ + (Ia). $[AIBN]_0 = 5 \cdot 10^{-3}$ M, temp. 70° C.

We suppose that when nonene–1 oxidation is inhibited with $(IIa)_{Ni}$, at the first stage the ligand is displaced from the coordination sphere of the metal. This suggestion is supported by the increasing optical density in the UV region at the absorption wavelength of ligand (Ia), as well as by the enhancement of the EPR signal, which is due to the greater intensity of the signal from the ligand as compared with that from the $(IIa)_{Ni}$ complex with uncoupled paramagnetic centers. The subsequent decline in curves 1 and 3 in Figure 3b is caused by consumption of (Ia). The acting concentration of the complex, reflected in the optical density of the solution at 258 nm, monotonously decreases with time (curve 2 in Figure 3b).

To check the suggestion about the appearance of transformed complex $(IIa)_{Cu}^{ox}$ during nonene-1 oxidation, free ligand (Ia) has been introduced into the reaction mixture. As can be seen in Table 2, combinations of $(IIa)_{Cu}$ and (Ia) produce a synergistic effect ($\tau_{mix} > \tau_{II} + \tau_{I}$) on the nonene–1 oxidation, which is in all probability due to displacement of oxidized ligand from the metal complex sphere, similar to what has been observed with copper dithiocarbamate [10]. This suggestion is also favored by the fact that in this case EPR does not register any $(IIa)_{Cu}^{ox}$ formation, the signal intensity decreasing monotonously (curve 4 in Figure 3a).

The activity of $(IIa)_{Ni}$ as an oxidation inhibitor does not appreciably change upon addition of (Ia) (Table 2).

Table 2. Effect of added (Ia) on the lag period τ in nonene-1 oxidation in the presence of $(IIa)_M$. $[AIBN]_0 = 5 \cdot 10^{-3}$M; $(IIa)_{Cu}$ $7.3 \cdot 10^{-4}$M; $(IIa)_{Ni}$ $6 \cdot 10^{-4}$M.

	τ, min	
Concentration of (Ia), M	$(IIa)_{Cu}$	$(IIa)_{Ni}$
None	5	12
$2.5 \cdot 10^{-4}$	10	nd
$5.0 \cdot 10^{-4}$	16	13
$1.0 \cdot 10^{-3}$	25	20
$1.5 \cdot 10^{-3}$	40	18

(nd), not determined

Since metal complexes $(IIa)_M$ can quite readily react with free alkyl radicals, they can be expected to be efficient inhibitors of radical polymerization of unsaturated compounds. Styrene polymerization initiated with AIBN has been chosen as a model reaction for studying the inhibitory activity of $(II)_M$ towards radical polymerization. The polymerization process was monitored dilatometrically by the decrease in solution volume ($-\Delta V/V_0$).

As can be seen Figure 4, introduction of a stable nitroxyl radical (Ia) results in the appearance of a significant lag period in the kinetic curve of the solution volume. The polymerization rate after the lag period coincides with that of nitiated polymerization in the absence of an inhibitor.

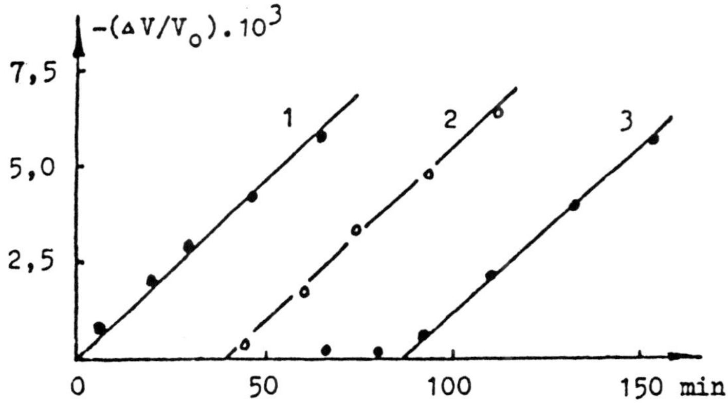

Figure 4. Kinetic curves of the changes in solution volume ($-\Delta V/V_0$) during radical polymerization of styrene: (1) control, (2) with $3.6 \cdot 10^{-4}$ M (Ia), (3) with $3.6 \cdot 10^{-4}$ M $(IIa)_{Cu}$. $[AIBN]_0 = 1.1 \cdot 10^{-2}$ M, temp. 60° C.

Table 3. Lag periods (τ) and inhibition coefficients (*f*) observed for initiated polymerizaion of styrene in the presence of (I) and $(II)_M$.
$[AIBN\}_0 = 1.1 \cdot 10^{-2}$ M, 60° C; inhibitor concn. 3.6 10^{-4} M.

Inhibitor	τ, min	*f*
(Ia)	41	0.88
(Ib)	49	1.0
(Ic)	52	1.1
$(IIa)_{Ni}$	110	2.1
$(II)_{Ni}$	98	2.4
$(IIa)_{Cu}$	88	1.9

The lag periods in styrene polymerization caused by $(IIa)_{Cu}$, $(IIa)_{Ni}$, or $(IIb)_{Ni}$ exceed twofold those for (Ia) and (Ib) at the same molar concentrations. The formal values of the stoichiometric inhibition coefficients calculated as above are given in Table 3.

The twice longer lag induced by the metal complexes appears to be due to the coordination of two ligands to the metal, and allows a conclusion that inhibition by the complexes takes place owing to the interaction of free radicals with the NO paramagnetic centers whose activity does not change upon coordination of the anions to copper or nickel.

$$M - NO + R \rightarrow M - NOR$$

It is noteworthy that, in contrast to the inhibition of nonene–1 oxidation, in the radical polymerization processes the formation of complexes of (I) with nickel(2+) or copper (2+) does not enhance their inhibitory activity.

It can be supposed that, besides the reactions of $(II)_M$ with free alkyl radicals proceeding at the NO center, an important part in the oxidative processes is played by "secondary" reactions including the interaction of the inhibitor conversion products M–NOR with peroxide radicals RO_2. Such interaction may involve the H—C bond in the α position with respect to the carbonyl group, which must result in a significant influence of steric factors and in the lack of correlation between the $E_{1/2}$ and the oxidation rate, as observed in the experiment. Inhibition of oxidation can also be substantially contributed to by the process of peroxide decomposition through both radical and molecular pathways.

3. Conclusions

(1) The inhibitory activity of nitroxyl radicals of various structure (RNO) towards oxidation of nonene–1 is substantially enhanced upon coordination of the anions to copper or nickel in $M(RNO)_2$. The inhibitory activity of the complexes depends on the nature of metal and ligand.

(2) It has been found that RNO markedly retard styrene polymerization; the inhibitory activity of the radicals in the polymerization process does not change upon coordination to copper or nickel.

4. REFERENCES

1. **A. L. Buchachenko and A. M. Vasserman**, "Stabilnye radicaly" (Stable Radicals), Moscow: Khimiya, 1973, 375 pp., Russian.
2. **A. N. Zverev, V. G. Vinogradova and Z. K. Maizus, Izvestiya,** *AN SSSR, ser. khim.*, No. 11, 1980, pp. 2479–2483, Russian.
3. **G. A. Kovtun, D. L. Lysenko and I. I. Moiseev**, *Koordinatsionnaya Khimiya*, Vol. 8, No. 11, 1982, pp. 1497–1502, Russian.
4. **V. N. Vetchinkina, I. P. Skibida and Z. K. Maizus**, *Izvestiya AN SSSR, ser. khim.*, No. 5, 1977, pp. 1008–1013, Russian.
5. **S. V. Larionov, V. I. Ovcharenko, V. N. Kirichenko, V. K. Mokhosoeva and L. B. Volodarskii**, *Izvestiya AN SSSR, ser. khim.*, No. 1, 1982, pp. 14–18, Russian.
6. **N. M. Emanuel, E. T. Denisov and Z. K. Maizus**, "Tsepnye reaktsii okisleniya uglevodorodov v zhidkoĭ faze" (Chain Reaction of Hydrocarbon Liquid-Phase Oxidation), Moscow: Nauka, 1964, 253 pp., Russian.
7. **G. P. Gladyshev and K. M. Gribov**, "Polimerizatsiya pri glubokikh stadiyakh prevrashcheniya i metody ee issledovaniya" (Polymerization at Deep Conversion Stages and Methods of Its Study), Alma-Ata: Nauka, 1968, 77 pp., Russian.
8. **V. I. Ovcharenko, L. A. Patrina, E. G. Biguslavskii, V. N. Ikorskii, V. B. Durasov, and S. V. Larionov**, Zhurnal Neorganicheskoĭ Khimii, Vol. 32, No. 5, 1987, pp. 1129–1133, Russian.
9. **A. N. Zverev, andV. G. Vinogradova**, *Izvestiya AN SSSR, ser. khim.*, No. 11, 1973, pp. 2440–2444, Russian.
10. **A. N. Zverev and V. G. Vinogradova**, Neftekhimiya, Vol. 22, No. 4, 1982, pp. 483–487, Russian.

Formation of the Reaction Front in Turbulent Flows

S.K. Minsker
Bashkirian State University, 32 Frunze str., 45076 Ufa, Russia

A.A. Konopljov
Institute of Pure Chemical Reagents, 3 Bogorodsky Val str., 10758 Moscow, Russia

T.V. Golubeva and Y.A. Prochkhan
Bashkirian State University, 32 Frunze str., 45076 Ufa, Russia

V.Z. Kompanietch
Institute of Pure Chemical Reagents, 3 Bogorodsky Val str., 10758 Moscow, Russia

G.E. Zaikov and Al.Al. Berlin
Institute of Chemical Physics of the USSR Academy of Sciences 4 Kosygin str., 117334 Moscow, Russia

Abstract—The existence of three different macrostructure of the reaction front was experimentally discovered on the example of interactions between solvents KSCN and $FeCl_3$ with the formation of colored reaction product in turbulent flow in tubular reactor. The optimal correlation of reacting flow rates was found and also the geometry of the agents introduction zone and the construction of branch pipes, which is the consequence of process proceeding in the regime of quasi ideal displacement in turbulent reactor. The formation of plane rapid reaction front in turbulent flow was confirmed by mathematical calculations of the process in the frame of "$q - \varepsilon$" turbulence pattern.

1. NOMENCLATURE

C/C_O—Fraction of the total mass of the 1 agent in the reactor.
C_d—The empirical constant
d—Reactor diameter
H—Blend braking enthalpy
h—Distance between the 1 agent introduction point and the reaction front
Jj—j vector components of the flux density of mass and heat
L—Distance between the 2 pipe axis and the 1 agent introduction point
N_k—Components number
N_r—Reactions number
P—Pressure
q—Turbulent energy
r—Reactor radius
V_1—Linear rate of the 1 agent flow at the reactor entrance
V_2—Linear rate of the 2 agent flow at the reactor entrance
v—The flow linear rate in the reactor
W—The rate of chemical reaction
Z/d—Distance on the reactor axis, in gages
α—Correlation of the linear rates of the 1 and 2 agents
Υ—Mass concentration
ε—Turbular energy dissipation rate
μ—Effective turbular viscosity
ρ—Blend density
τ_{ij}—Viscous stress tensor
Re—Reynolds number

The process of the formation of three different macrostructures of the reaction front in turbulent flow was experimentally studied on the example of reaction between KSCN and $FeCl_3$ with the formation of colored reaction product in tubular reactor.

There were found the optimal correlations of the reaction flow rates, the structure of the branch pipe and the geometry of the agents introduction zone, which makes it possible to form the preassigned macrostructure of the reaction front in rapid chemical reactions.

The possibility of the formation of the reaction plane front in turbulent reactor was confirmed by the mathematical calculations with "$q = \varepsilon$" turbulence model.

The process of the agent mixing organization is playing an important role in the optimal proceeding of the chemical reactions.

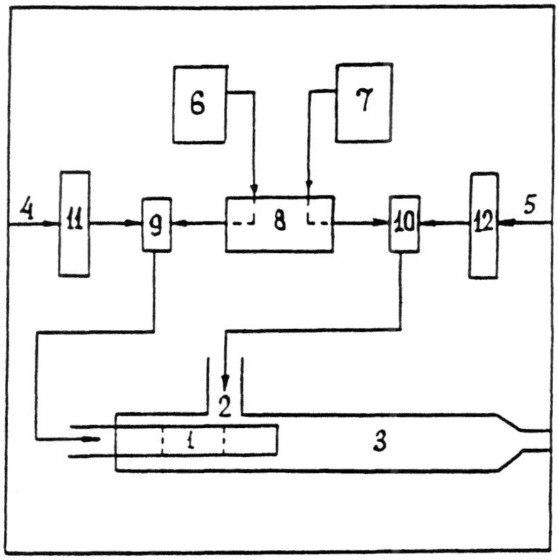

Figure 1. The scheme of installation for studying of the reaction front in turbulent flow. 1—branch pipe of the 1 flow introduction; 2—branch pipe of the agent 2 flow introduction; 3—reactor; 4,5—lines of water feed; 6,7—buffer capacity with agents; 8—peristaltic multicanal pump; 9,10—the place of water dilution; 11,12—rotameter.

This proved to be of present interest in the case of rapid chemical reactions, when the characteristic time of the proper chemical reaction is comparable or less than characteristic time of agents transfer.

There were found three macrokinetic regimes of the chemical process while studying the fast polymerization of isobutylene in turbulent flows with the help of mathematical simulation methods [1]. The found regulations proved to be general for other rapid chemical reactions such as chlorination and hydrochlorination of ethene, sulfuric alkilation of isoparaffins by isoolefins, oligomerization of piperylene in turbulent flows and so on..

In present paper we tried to study the formation of the front reaction macrostructure on the example of interactions between KSCN and $FeCl_3$ in turbulent flows on model plant in order to find out the conditions and reasons of formation of different fronts in rapid chemical processes (Figure 1).

There were used the branch pipes of three types: A—with the open end; B—with the radial openings; C—with the openings, placed in pairs or in a spiral. The expansion of the range of linear ranges of the first agent supply was provided by change of the pipe diameter (4–15 mm), change of the diameter (2–5 mm) and quantity (2–72) of the openings in the pipes B and C.

The distance (*L* mm) between the second pipe axis and the point of the first agent introduction was also changed. The agent introductions was realized by one of three methods: 1—L = 45 mm, 2—L = 0, 3 – L = – (45 mm). In the third case the first agent introduction point was placed before the second pipe.

The first flow rate (V_1) was changed from 20 to 700 sm/s and the second flow rate (V_2) was also changed from 10 to 40 sm/s. The flow rates were changed in the range 30–115 ml/s.

The experiment conditions provided the turbular character of the flow in reactor at all

preassigned agents introduction rates ($Re > (4 \pm 2)*10^3$).

Depending on the first pipe structure, geometry of the agents introduction zone and linear flow rates the formation of the reaction front structures of three types was observed.

F—reaction front in topochemical aspect is analogous to the gas burner flame or to the torch, which expands while moving off the first agent introduction point by reactor coordinate (Figure 2,a).

P—reaction front has the plane structure and is situated perpendicularly or at some angle to the reactor axis. Regime of the reactor work at the formation of the plane front can be defined as the regime of quasi-ideal displacement at the reaction process in turbulent flows.

The plane reaction front coordinate on the reactor axis may have both positive value relative to the first agent introduction point (Figure 2,b) and negative (Figure 2,c).

Z—the regime of the reactor work, where the strong back diffusion of the agents with the separation current takes place. This causes the distribution of the reaction product through the reactor zone, placed before the first agent introduction point (Figure 2,d) and the formation of stagnant zones.

The analysis of the results shows that for the formation of the concrete structure of the front in rapid chemical reactions it is important to know not only the absolute values of the flow rates but also their ratio ($V_1/V_2 = \alpha$). The types of the reaction structures depending on the correlation of linear rates of the agents flows in both flows yield; the structure of the first branch pipe and the geometry of the agents introduction zone are given in Table 1.

Table 1. Structure of the Reaction front depending on the method and agents introduction rate.

Agent introduction method	α	Type of the reaction front
	0.5–1.5	F
	1.5–2.0	mixed (F,P)
A–1	2.0–5.0	P
	5.0–8.0	mixed (P,Z)
	>8.0	Z
	<1.5	P
A–2	1.5–2.0	mixed (P,Z)
	>.2.0	Z
A–3	0.5–25	Z
B–1	0.5–60	P
	>60	Z
B–2	0.5–60	Z
B–3	0.5–60	P
C–1	0.5–20	P
	>20	Z
C–2	0.5–20	Z
C–3	0.5–10	P

The reaction front of F-type is clearly seen only in the case of A-1 method mixing, when the first pipe with open end is used (L = 45 mm). The reaction structure F is realized as a result of insufficiently developed turbulent flow.. Carrying out the rapid chemical reactions in such conditions is not effective, because of the formation of the agents overshoot zones and presence of the agents concentration gradient and temperatures in the reactor volume. This negatively influences both on the yield and on the quality of the reaction product.

Regime Z was observed for all selected structures of the 1 branch pipes. At L = 45 mm regime with the back diffusion is formed with the increasing of α values. At A–1 mixing method the α transitional value was 8; at C–1 and B–1 methods α = 20 and 60 correspondingly. At the agents introduction to one point (L = 0) the only one type of the reaction

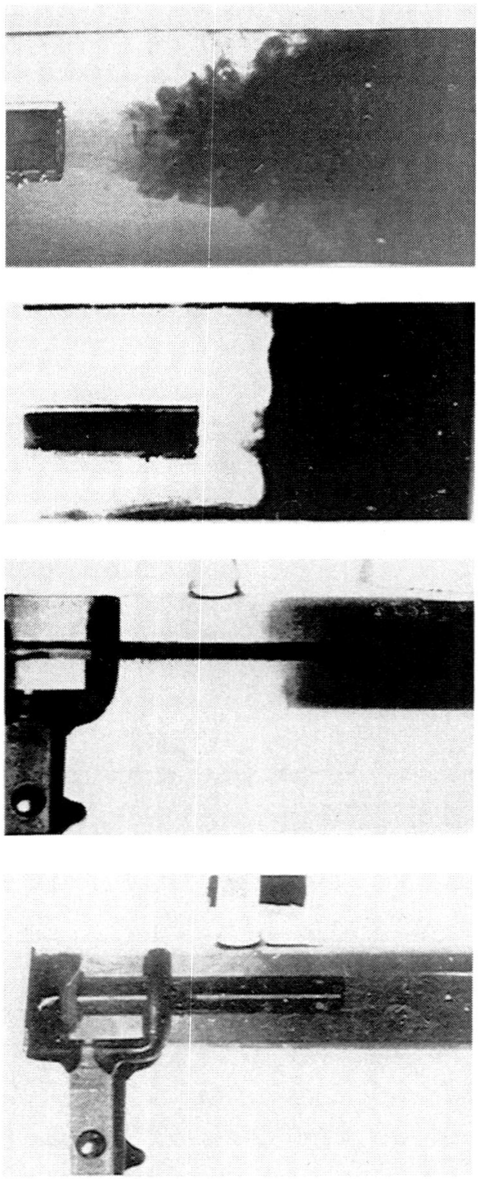

Figure 2 The reaction front at mixing of turbulent flows of KSCN and $FeCl_3$ in tubular reactor with the help of methods: a -A- 1, $\alpha = 1$; b- A-1, $\alpha = 3.5$;– c – B -1, $\alpha = 10$; d -B-1, $\alpha = 60$.

front at all mixing methods is macroscopic regime Z. In this case because of the separation currents there formed the stagnant zone in reactor. It is situated between the beginning of the turbulent reactor and the first introduction point. The presence of the stagnant zone was clearly observed in such experiments, where after the installation of the regime the agents supply to the mixes stopped, but the flow rates remained the same.

At providing the rapid chemical reactions in flowing turbular reactor in the regime Z the time of the presence of the part of the product in reaction zone increases. This causes the increase of by-products yield and the decrease of the process selectivity. Besides, the increasing time of the reaction product presence in stagnant zones at rapid processes of polymerization can promote the reaction of macromolecular sewing and the formation of insoluble polymer products which in particular were observed in model experiments at butylrubber and oligopiperylene receipt (the 1st flow—catalyst, the 2nd flow—monomer or monomer blend).

The most favorable regime for the providing of the rapid chemical reactions in turbulent flows is the quasi-ideal displacement at which the plane reaction front is formed (type P).

Regime P is realized thanks to the strong developed turbulence in the agents mixing zone and is characterized by homogeneous composition of the reaction flow along the reactor radius. This allows to get the polymeric product with the normal molecular-mass distribution at the reaction of isobutylene polymerization [1].

At the organization of the mixing process by A–1 method the consecutive increasing of α values cause the change of macrostructure of the reaction front from torch through the plane up to the Z type. The clear plane front is formed in the narrow a values range from 2 to 5. In many cases the reaction front of the types F and P is not situated near the first agent introduction zone but at different distances from it.

Thus at mixing of two turbulent flows by A–1 method the specific "dead zone" is formed and there is no turbular mixing of the flows in it. Depending on α values at a definite distance (h) from the first agent introduction point the fronts F and P are formed unevenly.

The dependence between the distance to the reaction front and a values is presented in Figure 3. The analytical expression of this dependence is: $h/d = 1.0 - 0.2^{*}\alpha$. The correlation factor is 0.93.

In the case of branch pipes with radial introduction of the 1 agent (types B and C) at mixing by B–1 and C–1 methods regime P is considerably more stable than that of the A–1 method.

As it is seen from Table 1 the plane structure of the reaction front is formed in wide range of α values.

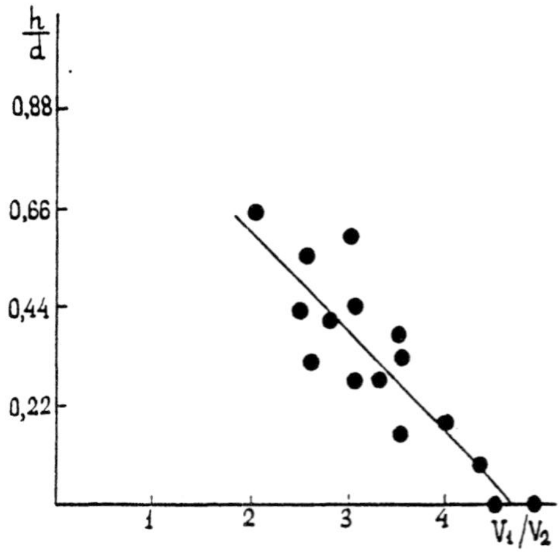

Figure 3. The dependence of distance till the reaction front on relation of linear rates of flows at mixing with the help of method A–1.

It is important to note that in these cases the mixing front boundary always has the negative coordinate on the reactor axis relative to the first agent introduction point.

There is skidding of the reaction front because of the back diffusion and like in A–1 case the increasing of values at this boundary moves to the opposite side of the downflow in the reactor. Thus the providing of the chemical process by B–1 and C–1 methods on one side ensures the balanced distribution of the agents in mixing zone and absence of the stagnant zone in the reactor, and on the other side it conducts the constant contact of the reaction product both with the pipe surface in the 1 agent introduction zone and with the 1 agent.

At mixing of the flows by B–3 and C–3 methods there observed the formation of the reaction plane front at all studied flow rates. The front boundary is situated at the distance (by downflow) from the first introduction zone. The specific feature of the mixing methods B–3 and C–3 is moving of the front boundary from the first introduction zone by downflow with the increasing of a values. Here we observe the reverse effect to that of the A–1, B–1, C–1 methods.

To understand the process of the formation of different macrostructures of the reaction front at providing of the rapid chemical reaction in turbulent flows it was made the mathematical simulation of the mixing process in the frames "q – ε" on the turbulence pattern. It allows to take into account the influence of hydrodynamical conditions of the current on the process of mixing and chemical transformation [2].Mixing method A–1 was chosen as a pattern. Geometrical sizes of the reactor and flows rates correspond to the experimental one.

For computing the turbulent currents there were used the equations for average pulsate values:

$$\partial(\rho * v_{ii})/\partial x_i = 0 \tag{1}$$

$$\partial(\rho * v_j * v_i)/\partial x_j - \partial(\tau_{ij}^t)/\partial x_j + \partial P/\partial x_i = 0 \tag{2}$$

$$\partial(\rho * v_j * H)/\partial x_j + \partial(J_{H_j})/\partial x_j = 0 \tag{3}$$

$$\partial(\rho * v_j * \Upsilon_s)/\partial x_j + \partial(J_{sj})/\partial x_j - \sum_{r=1}^{N_R} \dot{W}_{sr} = 0 \tag{4}$$

To closure the system of Equations (1–4) for average values supplied by the state equation $\rho = \rho$ (T) semiempirical two-parameter turbulence model was used. It contains of two supplement equations for the kinetic energy of pulsate movement "q" and average specific velocity of its dissipation "ε." The efficient turbulent viscosity is defined by the equation:

$$\mu_t = C_{D*} q^3/\varepsilon$$

Several results are summed up in Figure 4. The turbulent energy of the flow at the 1 agent yield in the center of the steam has the maximum value. At the distance r/d=0,2 from the reactor wall q value is 5–10% of the flow energy in the center (Figure 4,a, curve 1). Besides at a distance near one gage from the 1 agent introduction point the considerable decrease both of the absolute value of turbulent energy and its gradient on reactor radius is observed (Figure 4,a, curve 2). The efficiency of mixing in reactor considerably decreases at a distance which is more than the gage value relative to the 1 flow introduction point along the reactor axis. At the distance—0.5 Z/d from the 1 agent introduction point (Figure 4,b, curve 1) the steam preserves the nucleus but at Z/d>1 it decomposes with the formation of practically homogeneous mixed flow (curve 2). At low α values (<1.5) the profile of the reaction zone is like a "torch." Near the reactor walls there is a zone (r/d ≥ 0.4) in which the 1 agent is absent (Figure 4,b, curve 1). The increasing of α values causes the emergence of the 1 agent in the outlying district and the decrease of the gradient concentration along the reactor radius. (Figure 4,b, curve 3). Taking into account the experimental conditions the formation of the plane front is observed and this really takes place in the experiment on the model reactor (Table 1).

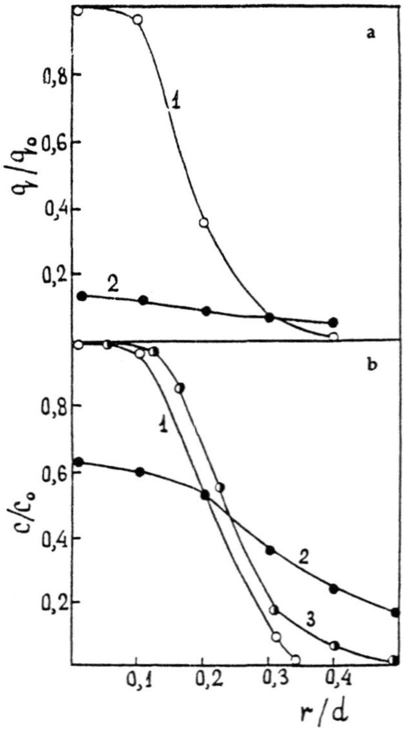

Figure 4. Distribution along the reactor radius: a—turbulent flow energy: a = 1.6; 1–Z/d = 3.1; 2–Z/d = 4.0; b—agent 1 (mass share): Z/d = 3.1 (1–b,3–b); Z/d = 4.0 (2–b); a= 1.0 (1–b,2–b); a = 2.2 (3–b).

2. REFERENCES

1. **Al.Al Berlin, K.S. Misker, Y.A. Prochukhan, M.M. Karpasas,** and **N.S. Enikolopyan,** *Macromolec. Comp., B.,* Vol. 28, No. 6, 1986, pp. 461–465.
2. **V.Z. Companietz, A.A. Ovsyannikov,** and **A.S. Polak,** Chemical Reactions in Turbulent Flows of Gas and Plasma, M., Science, 1979, pp. 242.

Radical Oxidation in Non-Equilibrium Glass *N* - Butanol–*D*10

S.V. Vasenkov and V.A. Tolkatchev
Novosibirsk State University, 630090, Novosibirsk,
Russia Institute of Chemical Kinetics and Combustion,
630090,Novosibirsk, Russia

Abstract—The kinetics of radical oxidation and the influence of the preliminary annealing of matrix on this kinetics at different temperature were studied in the glassy *n*-butanol-d_{10} matrix using ESR technique. The oxidation kinetics was described in terms of a polychromatic approach allowing for the fact that in the temperature range 90–100 K the relaxation processes occur that decrease the rate constant in each ensemble according to one and the same law. It was shown that there are two temperature regions of thermal relaxation in *n*-C_4D_9OD matrix. The data obtained indicate that the matrix contains two types of regions: with more porous and more dense packings. Radicals in this case are mainly distributed in the regions with a more porous packing.

1. Introduction

Hypothesis on rate constant distribution has often been used to describe the formal kinetics of solid phase chemical reactions [1,2] which is confirmed by serious experimental evidence [1–4]. However, it fails to account for the kinetics of free radical oxidation by molecular oxygen $R + O_2 \rightarrow RO_2$ for a series of low-molecular glasses [5–10]. According to [5–10] the process can be interpreted in the framework of the time-dependent rate constant formalism:

$$\frac{d[R]}{dt} = -k(t)[R][O_2] \quad (1)$$

where $k(t)$ decreases with increasing t and is dependent on the matrix and the way of its preliminary treatment.

The influence of the preliminary thermal annealing of a number of matrices (i.e. annealing before radical generation) at reaction temperature (87K, 90K) on the kinetics obeying (1) has been studied in References [9,11]. It was shown that in all cases the preliminary annealing at reaction temperature slows down its rate. The purpose of the present report is to describe in detail the influence of annealing (using one of these matrices) on the kinetic process at a quantitative level and to study the reasons for such an unusual kinetic behavior.

The study of annealing mechanism and its role in solid phase reactions has its own essence. There are many papers demonstrating that solid phase reactions can be subjected to both mechanical matrix nonequilibrium and the relaxation processes therein. Thus, e.g. the low-temperature explosive reactions of saturated hydrocarbon halogenation, olefin hydrobromination, and polymerization occur in the mechanically nonequilibrium media only. The structural relaxation processes in these matrices make the character of these reactions unusually wavy ([12–18] and references herein).

It is commonly assumed that the annealing of low-molecular matrices below melting point for crystalline systems and below the glass transition point for glassy ones just slows down the kinetic process [12,19,22]. This is assigned to the reduction of the free volume of the system. The influence of annealing on polymethylmethacrylate is caused not by a simple reduction of the system free volume which slows down chemical transformations [23]. The kinetics can be explained by a detailed description of matrix relaxation at a microscopic level taking into account the changes in the distribution function of low density clusters [23]. The present paper testifies that a simple approach fails to describe the influence of annealing on the kinetics of free radical oxidation in low-molecular substance—n-C_4D_9OD. Annealing in this case leads to fairly complex phenomena that either accelerate or slow down the chemical processes.

2. EXPERIMENTAL

A frozen low-molecular n-C_4D_9OD, supplied by "Isotope" and containing 96% of deuterated molecules was used as the matrix. 2.5 vol. % benzyl chloride of a "pure" grade and 10^{-3} mol. % diphenylamine (DPA) were added to alcohol. Oxygen of a "pure" grade was purified from water vapors by passing through a coil below 200K. Benzyl radicals were generated at 77K using dissociative capture of electrons formed either on photoionization of DPA or on radiolysis on a γ-source ^{60}Co of 0.67 Mrad/hr. The γ-irradiation dose was 0.17 Mrad. Photolysis was performed with the help of a DRSch-100 mercury high-pressure lamp with a water filter. Irradiation time was 2 minutes. The samples were saturated with oxygen at room temperature by two methods. In the first case a constant oxygen pressure (P) was created in the ampoule with *n*-butanol-d_{10} at room temperature. The liquid was then stirred during the time of gas dissolution. In the second case (with $P > 12$ atm.) the samples were saturated with oxygen by freezing oxygen in the ampoule with C_4D_9OD from a calibrated volume of a vacuum system. The ampoule was then sealed, unfrozen, and kept at room temperature for a certain time. After saturation the ampoules were frozen in liquid nitrogen.

Oxidation and annealing of samples were carried out in a thermostabilized nitrogen jet to within ±0.3K and in liquid oxygen at 90K. Radical concentration was measured by ESR method at 77K, on a RE 1306 ESR spectrometer. The mw field power was chosen so that the signal was far from saturation. Oxygen solubility in our system obeys the Henry law and is $\beta = (4.4 \pm 0.8) \times 10^{18}$ cm^{-3} atm^{-1} [7,24]. Oxygen concentration in samples was varied from 4.3×10^{17} cm^{-3} to 3.5×1020 cm^{-3}. The radical concentration was in the range 0.5×10^{17} cm^{-3} to 2×10^{17} cm^{-3}. The condition of reaction pseudo-mono-molecularity $\left([R]^{-}[O_2]\right)$ held for all cases. Kinetics was measured in both relaxed and nonrelaxed samples. The matrix was relaxed before radical generation using either thermal annealing or mechanical destruction of samples with a metallic needle at 77K as described in [25].

Photoionization was used to obtain radicals at saturated pressures $P < 10$ atm. At large pressures, probably, due to a strong quenching of the excited states of DPA molecules by oxygen, this method provides a low yield of radicals. Therefore at $P > 10$ atm. the γ-irradiated samples were used. The tests performed at $P = 5$ and 10 atm. show that the two types of radical generation cause no difference in their kinetic behavior.

3. Results and Discussion

3.1. Relaxation Regions

Experimental conditions were similar to those described in [5–11]. The ESR spectrum obtained on either photolysis or γ-radiolysis at 77K coincides with that given in the literature [26] for benzyl radicals. With concentrations $[O_2] > 3.5 \times 10^{19}$ cm^{-3} the sample spectra exhibit the lines belonging to peroxide radicals and O_2- ones. When temperature increases to 90 or 98K the spectrum of benzyl radicals transforms to the peroxide ones.

It was found that the decrease in benzyl radical concentration in degassed samples during annealing at reaction temperature (90K and 98K) and the maximum reaction time was below 15%. According to these and experimental data in References [7–9] it is concluded that in our experimental conditions in the presence of O_2 in the n-C_4D_9OD matrix the benzyl radicals decay via the process $C_7H_7 + O_2 \rightarrow C_7H_7O_2$. The contribution of other decay channels is either absent or negligible.

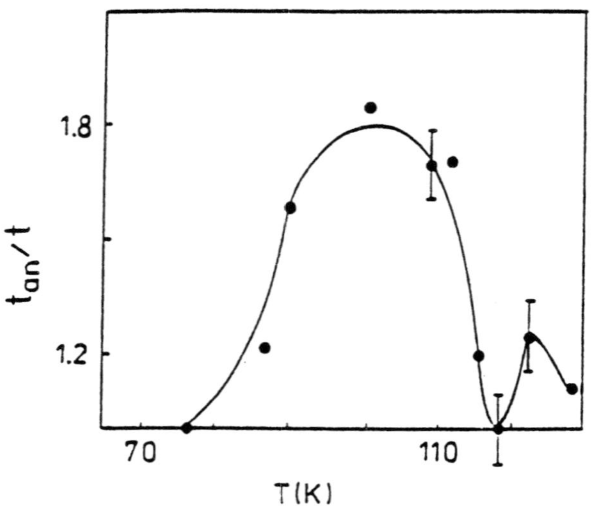

Figure 1. The dependence of the relation of reaction times at which the transformation depth of 68% is reached in the annealed and not annealed samples (tan/t) on annealing temperature. The duration of annealing is 120 min.

Consider now the influence of the preliminary annealing of samples with $[O_2] = 3.5 \times 10^{19}$ cm^{-3} at different temperature on the oxidation kinetics at 90K. Figure 1 depicts the dependence of the relation between reaction times at which the transformation depth of 68% ($\log[R]/[R0] = -0.5$) is reached in annealed and not annealed samples on the annealing temperature.

Figure 1 shows that as the annealing temperature increases from T = 87K to T = 100K, the oxidation reaction rate (at 90K) gradually slows down and the maximum effect is observed during annealing near 100K. At higher temperature this effect decreases and at $T = 117 \pm 2$K annealing has no effect on the reaction rate at 90K. In addition, when the samples are kept at 117K the influence of annealing at 90K on the oxidation kinetics is the same as that in the not annealed samples. When annealing is carried out at T = 122K the process rate starts to slow down again.

From Figure 1 it follows that the effect of reaction slowing down due to annealing at T = 122K is quite low. Nevertheless, in contrast to T = 117K, after this procedure the proceeding annealing at T = 90K has a minor effect on the reaction kinetics (see Figures 2a and 2b). Indeed, the annealing at 90K for 66.5 hours alters the kinetic curve slope in the samples preliminarily annealed at 122K by no more than 20%. However, in the samples that were not subjected to a high-temperature annealing (122K) this operation decreases the kinetic curve slope more than 1.6 times (Figures 2a,b). Note that the matrix relaxation by destruction before radical generation has the same effect on the reaction kinetics as the annealing at T = 122K (Figure 2b).

1,2—in nonrelaxed samples with oxygen concentration of 3.5×10^{19} cm^{-3} and 3.5×10^{18} cm–3, respectively; 3,4—in the samples with oxygen concentration of 3.5×10^{19} cm^{-3} relaxed by destruction at T = 77K and by 2 hours annealing at T = 122K, respectively; 5,6—in the samples with oxygen concentration of $3.5 \times$ 1018 cm^{-3} relaxed by annealing at T = 122K and destruction at T = 77K; 7—in the samples with the mixture of $[O_2 = 3.5 \times 10^{18}$ cm^{-3} and $[N_2] = 1.9 \times$ 1019 cm^{-3}, relaxed by destruction at 77K.

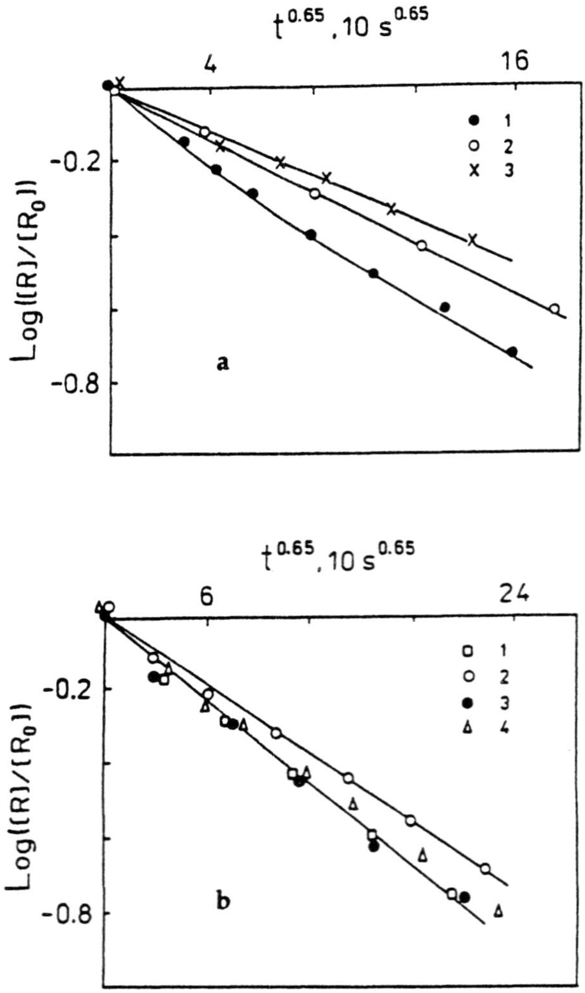

Figure 2. The kinetics of radical oxidation at T = 90K. Oxygen concentration is 3.5×10^{19} cm^{-3}.

a) Annealing at T = 90 K. The duration of annealing: 1—0 hours; 2—20.5 hours; 3—47 hours.

b) 1—annealing at T = 120 K for 2 hours; 2—annealing at T = 122K for 2 hours. and then at T = 90K for 66.5 hours; 3—mechanical destruction at T = 77K, then annealing at T = 90 K for 41 hours.

It appeared that the influence of annealing at T = 122K and the mechanical destruction of matrix on the oxidation process depend on oxygen concentration. As has been mentioned the annealing at 122K and the mechanical matrix destruction slow down the reaction rate at T = 90K in the samples with oxygen concentration $[O_2] = 3.5 \times 10^{19}$ cm^{-3} (compare curves 1 and 3,4 in Figure 3). At the same time the identical effect dramatically increases the reaction rate in the samples with oxygen concentration of 3.5×10^{18} cm^{-3} (curves 2 and 5,6, Figure 3). Thus the annealing at T = 122K and the destruction at 77K slow down the reaction rate at 90K in the samples with a high oxygen concentration and increase it in the samples with low oxygen concentration. At the same time the annealing of samples at 90K slows down the oxidation rate at 90K in the samples with both low and high oxygen concentrations. According to the above experimental data there are two temperature regions of thermal relaxation for our matrix.

It might be assumed that one of the reasons that the annealing at 122K and the mechanical destruction of matrix on the oxidation process depend on oxygen concentration in samples is the dependence of matrix relaxation properties on the amount of dissolved gas. This assumption was verified by the following experiment. We have prepared a matrix with the mixture of $[O_2] = 3.5 \times 10^{18}$ cm^{-3} and $[N_2] = 1.9 \times 10^{19}$ cm^{-3}. After mechanical destruction the oxidation kinetics in such matrix is the same as in the sample without N_2 (Figure 3). Thus, it might be concluded that the gas dissolved in the matrix has no influence on its relaxation properties.

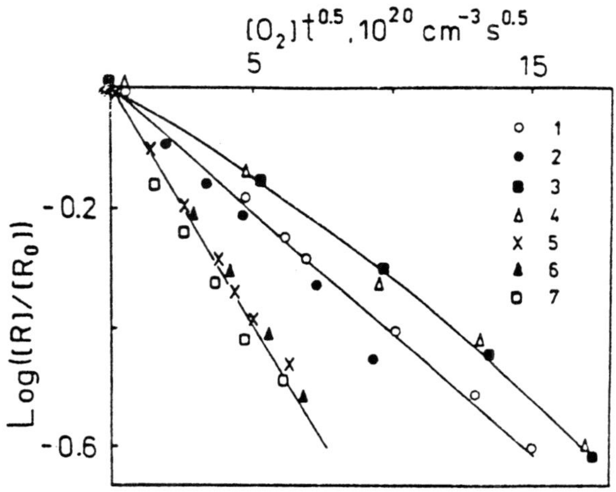

Figure 3. The kinetics of radical oxidation at T = 90 K:

The influence of annealing at 117K on the oxidation kinetics in the samples with low oxygen concentration was checked by the following experiments. When the oxygen concentration was $[O_2] = 3.5 \times 10^{18}$ cm^{-3} the sample annealed at T = 90K for 22 days was then annealed at T = 117K for 2 hours. The kinetics in this matrix coincided with that in the not annealed sample (Figure 4) (as for high oxygen concentrations). The figure testifies that the oxidation kinetics in the same sample annealed at T = 122K for 2 hours. after annealing at T = 90K for 22 days coincides with the kinetics in the sample with $[O_2]$ = 3.5×10^{18} cm^{-3} annealed only at T = 122K for 2 hours. These experimental data show that two relaxation regions and the losses of memory at 117K for low-temperature relaxation hold for any oxygen concentration.

Let us consider the reason for the increase in oxidation rate due to relaxation at low oxygen concentrations. The matrix relaxation causes, as a rule, the minimization of its free energy and, as a consequence, slows down the molecule mobility. It is impossible to understand the increase in the oxidation rate in relaxed samples from this viewpoint. It might be assumed that in the non-relaxed matrices there are such deep oxygen traps that the molecules trapped during freezing cannot participate in diffusion. If matrix relaxation by annealing at 122K destroys these traps, this sets free an additional portion of oxygen and accelerates the reaction.

The literature data are available confirming a possible existence of traps in glassy polymeric films where gas loses its ability to participate in diffusion [27–30]. Meares was the first to draw such a conclusion [27] to account for the measured values of the coefficients of diffusion, permeability, and solubility of different gases in polyvinyl acetate films. In 1958 Barrer *et al.* proposed a dual-sorption model [28] to explain the non-linearity of sorption isotherms in ethyl cellulose films. According to this model the solubility of the gas which does not participate in diffusion through the film, i.e. is trapped, obeys the Langmuir isotherm. The solubility in the other part of the matrix was described in terms of the conventional Henry law. At present there is a sufficient number of literature data confirming the applicability of dual-sorption model for describing the solubility and transport of various gases in glassy polymeric films (see, e.g. [29]).

Using these data we assume that our matrix also has a certain number of such traps and, as a consequence, the oxygen molecules that do not participate in diffusion. The presence of such traps will have no influence on the oxidation rate if the total oxygen concentration in the matrix is much higher than that of traps. In this case the process must slow down after annealing which explains the slowing down of the oxidation rate at high O_2 concentration.

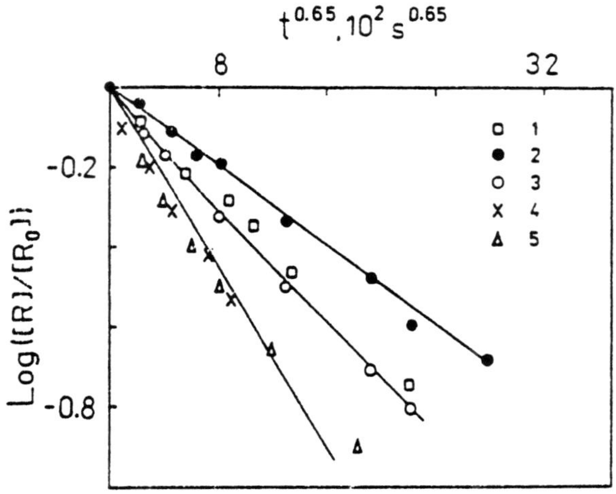

Figure 4. The kinetics of radical oxidation at $T = 90K$. Oxygen concentration is 3.5×10^{18} cm^{-3}. 1—without annealing; 2—annealing at T = 90K for 22 days; 3—annealing at $T = 90K$ for 22 days and then for 2 hours at $T = 117K$; 4—annealing at $T = 122K$ for 2 hours; 5—annealing at $T = 90K$ for 22 days and then at $T = 122K$ for 2 hours.

Consider now a possible origin of two temperature regions of thermal relaxation in our matrix. Two hypotheses can be used to interpret a low-temperature annealing (i.e. annealing at $T = (90–100K)$ in low molecular alcohol glasses. One of these is based on the concept of oxygen transport through the traps of the different depths without changing matrix properties. In the framework of this hypothesis the slowing down of oxygen diffusion during low-temperature annealing is related to the redistribution of oxygen molecules over deeper traps. The fact that the annealing at $T = 177 \pm 2K$ ceases to slow down the radical oxidation at 90K can be interpreted in the following way. It is assumed that at $T = 177K$ the trap depths start to vary with time with the frequency exceeding the mean frequency of oxygen molecule jumps. In this case at $T = 117K$ all the traps become equivalent, and no recapturing is observed. The freezing of such a matrix to 77K recovers the function of distribution over trap depths.

A high-temperature annealing and mechanical destruction of the matrix causes changes in its structure properties. This can lead to the narrowing of distribution over trap depths in the matrix either frozen to 77K after the high-temperature annealing or subjected to destruction at 77K. In other words, very "deep" and very "shallow" traps disappear. The distribution narrowing destroys the reasons for both recapturing (low-temperature relaxation) and the non-participating part of oxygen which explains the experiment.

The second hypothesis is assigned to the minimization of free matrix energy during low-temperature annealing due to matrix molecule motion. In order to explain the existence of two relaxation types, it is more natural to assume the existence of two regions with different molecule mobility for our matrix. More experimental data are now available in the literature testifying in favor of the regions with two types of molecule ordering in glasses (more porous and more dense regions) and two types of mobility in these regions (see References in [31,32]). Let us assume that the radicals are mainly formed in the regions with a more porous structure. In this case the low-temperature annealing (90 ÷ 100K) can be assigned to the relaxation of porous regions to some quasi—equilibrium state. With $T = 117 \pm 2$K the porous regions transfer from a solid to quasi—liquid state. Consequently, the memory about the low-temperature relaxation is lost. At $T = 122$K the matrix molecule mobility is unfrozen in the regions with dense packing which enables the overall matrix relaxation to the equilibrium state and provides the irreversible destruction of oxygen traps. A mechanical destruction of the sample gives the same results.

In order to choose between these two interpretations it is necessary to consider our kinetic data.

3.2. Kinetics Peculiarities

It was shown [8,9] that during the radical oxidation in *n*-butanol-d_{10} in excess oxygen the logarithm of radical concentration must be proportional to ta, where a approaches 0.5 in the not annealed and 0.65 in the preliminarily annealed at 87K samples. This corresponds to our data (Figures 2,4–6). In this case $\alpha = 0.65$ remains the same during annealing at higher temperatures.

The formal kinetics has usually been described using a polychromatic approach. According to this approach all reacting particles can be divided into the groups reacting with their own rate constant [1,2]. In the framework of this theory the kinetics of radical oxidation at $[R]^{-}[O_2]$must obey the Equation (1,2):

$$\frac{[R]}{[R_0]} = \int g(i)\exp\left(-k_i[O_2]t\right)di \tag{2}$$

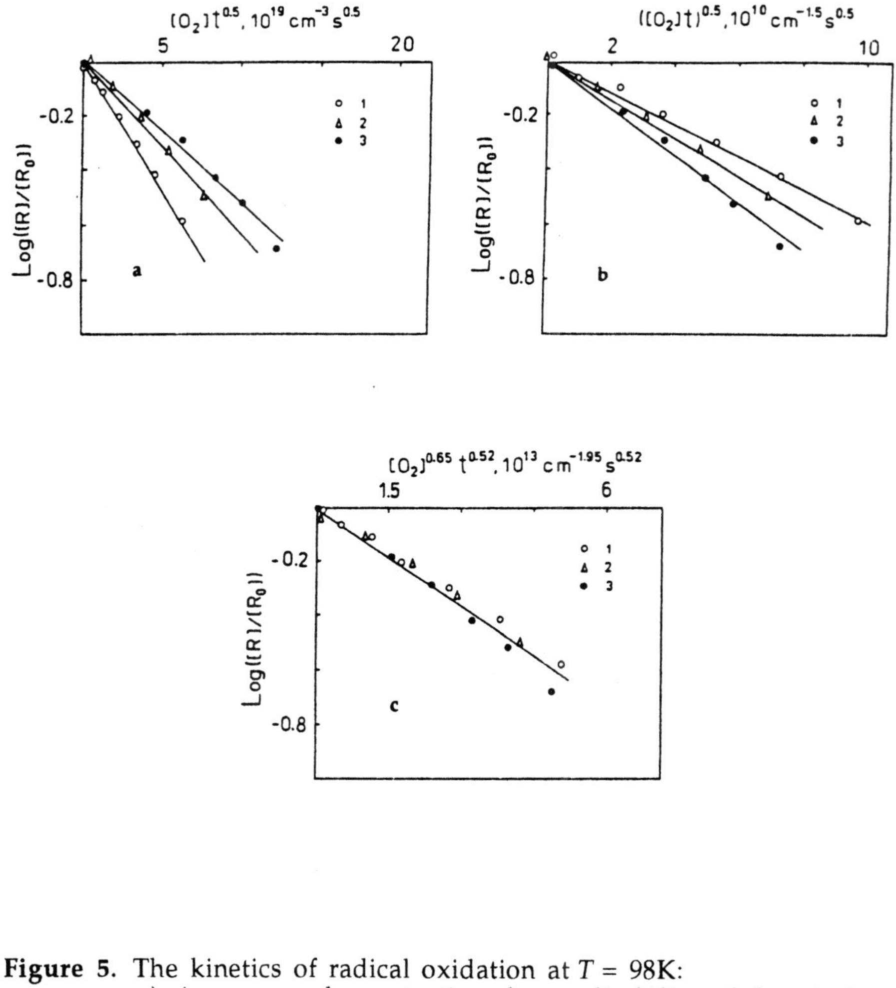

Figure 5. The kinetics of radical oxidation at T = 98K:
a) in anamorphoses testing the applicability of description (1); b) in anamorphoses testing the applicability of description (2); c) in anamorphoses testing the applicability of law (5). Oxygen concentration: 1—0.44 $\times 10^{18}$ cm^{-3}; 2—1.32 $\times 10^{18}$ cm^{-3}; 3—3.08 $\times 10^{18}$ cm^{-3}.

where k_i is the rate constant in the i-th group; $g(i)$ is the weight of the i-th group. Thus a transformation depth is determined by the parameter $[O_2]t$ [5]. Therefore when $\ln([R]/[R_0]) \propto t^{\alpha}$ we obtain:

$$\frac{[R]}{[R_0]} = \exp\left(-K([O_2]t)^{\alpha}\right) \quad (3)$$

The criterion for the applicability of a polychromatic description in this case is the coincidence of the kinetic curves in anamorphoses:

$$\log([R]/[R_0]) \div ([O_2]t)^{\alpha}$$

Another method to describe the reaction kinetics in a solid phase is the formalism of time-dependent rate constant (1). In our experimental conditions the applicability of this description is verified by checking the coincidence of the kinetic curves with different oxygen concentration in anamorphoses [5–7]:

$$\log([R]/[R_0]) \div [O_2]t^{\alpha}$$

The formulae show that the difference in the models leads to the different dependence of the process kinetics on oxygen concentration. When oxygen concentration varies (e.g. decreases), the transformation depth, in the first case, at a given t varies to a lesser extent than in the second case because the process in the latter case slows down due to the decrease in O_2 concentration with further decrease of the reaction rate constant.

Since the criterion used to differentiate the models is the oxygen dependence, it is important to operate within the range of O_2 concentrations where it is slightly distorted due to the capture of oxygen by deep traps. It might be expected that at low O_2 concentrations the fraction of gas molecules captured by any traps will vary in proportion to the total oxygen concentration in the sample. In this case the fact that some part of oxygen does not participate in diffusion will have no influence on the kinetic law of oxidation.

Figure 5 depicts the kinetics of radical oxidation at T = 98K and $[O_2]$ = (0.44 ± 3.06) × 10^{18} cm^{-3} in coordinates testing the applicability of descriptions (1) and (2). The figure shows that this kinetics is not described by laws (1) and (2). It appeared that the formally kinetic curves in these samples can be described by the law (Figure 5c):

$$\frac{[R]}{[R_0]} = \exp\left(-K[O_2]^{0.65} t^{0.52}\right) \tag{5}$$

This law means that with decreasing oxygen concentration the rate decreases faster than during the rate constant distribution. Thus it might be assumed that in the given system the kinetic nonexponentiality, on the one hand, is related to the rate constant distribution and, on the other hand, these constants vary with keeping matrix at reaction temperature.

In order to check this assumption we have measured the kinetic curves in the sample annealed at reaction temperature for 13 hours before radical generation. This period is 2.3 times as long as the maximum reaction time in the not annealed samples. From Figure 6 it follows that the kinetics in the annealed samples obeys the law:

$$\frac{[R]}{[R_0]} = \exp\left(-K_1\left([O_2]t\right)^{0.65}\right) \tag{6}$$

i.e. is described in the framework of a polychromatic model. This is in fair agreement with the above hypothesis. In addition, from (6) it is clear that convolution (2) in the given system is well approximated by Equation (3) with $a = 0.65$.

We have tried to take into account the influence of annealing on the rate constant in terms of the Kohlrausch law:

$$k_i'(t,\tau) = 0.8a_i(t+\tau)^{-0.2} \tag{7}$$

where a_i is the constant; k_i' is the rate constant in the i-th group; t is the reaction time; τ is the duration of the preliminary annealing at reaction temperature. Thus we have assumed that annealing of matrix at reaction temperature has the same effect on the reaction rate constants in all groups. Consider now to what extent this hypothesis corresponds to experiment.

When $\tau = 0$ for one group we can write:

$$\left([R]/[R_0]\right)_i = \exp\left(-a_i[O_2]t^{0.8}\right)$$

Using (6), for the whole of matrix we obtain:

$$\frac{[R]}{[R_0]} = \exp\left(-K_2\left([O_2]t^{0.8}\right)^{0.65}\right) \tag{8}$$

It is readily seen that (8) coincides with (5). If $\tau \neq 0$, the kinetic law for one of the groups is of the form:

$$\left(\frac{[\mathrm{R}]}{[\mathrm{R}_0]}\right)_i = \exp\left[-\frac{0.8a_i[\mathrm{O}_2]}{\tau^{0.2}}\int_0^t \frac{1}{(t'/\tau+1)^{0.2}}dt'\right]$$

With $t < \tau$ one can neglect the relaxation t'/τ in the denominator of the subintegral expression. For this case we have:

$$([\mathrm{R}]/[\mathrm{R}_0])_i = \exp\left(-0.8a_i[\mathrm{O}_2]t\tau^{-0.2}\right)$$

Performing the convolution over all zones we obtain:

$$\frac{[\mathrm{R}]}{[\mathrm{R}_0]} = \exp\left(-K_2\tau^{-0.13}\left(0.8t[\mathrm{O}_2]\right)^{0.65}\right) \tag{9}$$

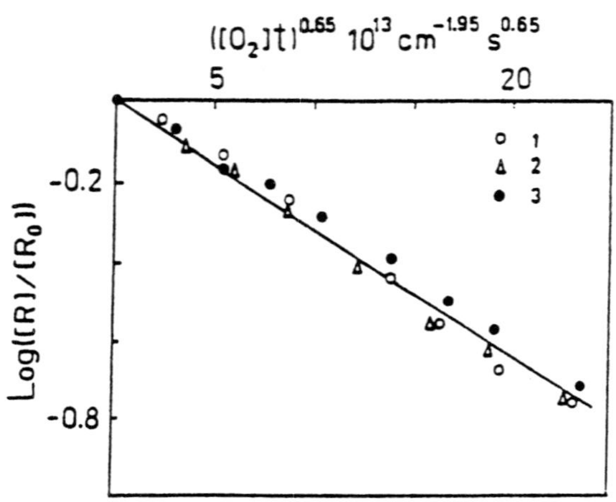

Figure 6. The kinetics of radical oxidation at T = 98K in the samples preliminarily annealed at T = 98K for 13 hours in the anamorphoses testing the applicability of polychromatic description. Oxygen concentration: 1—0.44 × 10^{18} cm^{-3}; 2—1.32 × 10^{18} cm^{-3}; 3—3.08 × 10^{18} cm^{-3}.

From (6) and (9) we deduce:

$$0.865 K_2 \tau^{-0.13} = K_1$$

or:

$$\log(K_1/0.865) = \log K_2 - 0.13 \log \tau \qquad (10)$$

Values K_1 and K_2 were determined from the slope of the kinetic curves in anamorphoses $\log([R]/[R_0]) \div ([O_2]t)^{0.65}$ and $\log([R]/[R_0]) \div t^{0.52}[O_2]^{0.65}$ in the annealed and not annealed samples, respectively.

Figure 7 depicts the dependence of $\log(K_1/0.865)$ on log τ. The dependence is observed to be satisfactorily described by Equation (10). Thus it is concluded that the oxidation kinetics with $[O_2] = (0.44 - 3.06) \times 10^{18} cm^{-3}$ is well described in the framework of the polychromatic approach allowing for the fact that at reaction temperature the relaxation processes whose influence on the reaction is described by expression (7) occur in each zone.

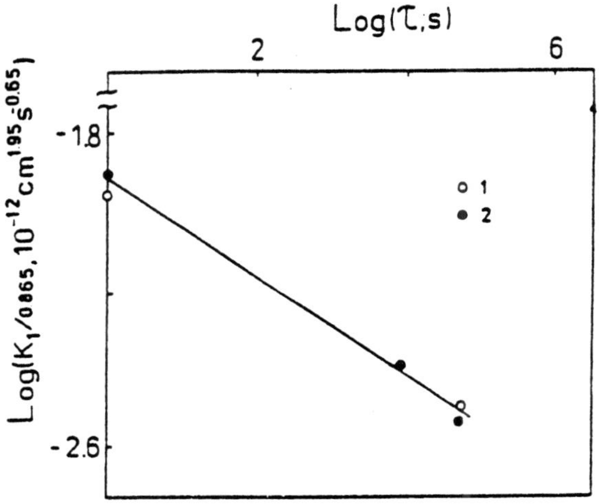

Figure 7. The dependence of logarithm k1 for the reaction at T = 98K on the logarithm of the duration of preliminary annealing at reaction temperature. Oxygen concentration: 1—0.44×10^{18} cm^{-3}; 2—3.08×10^{18} cm^{-3}. Points on the ordinate—log k_2 (10^{12} $cm^{1.95}$ $s_{-}0.52$). The slope of the straightline is 0.13.

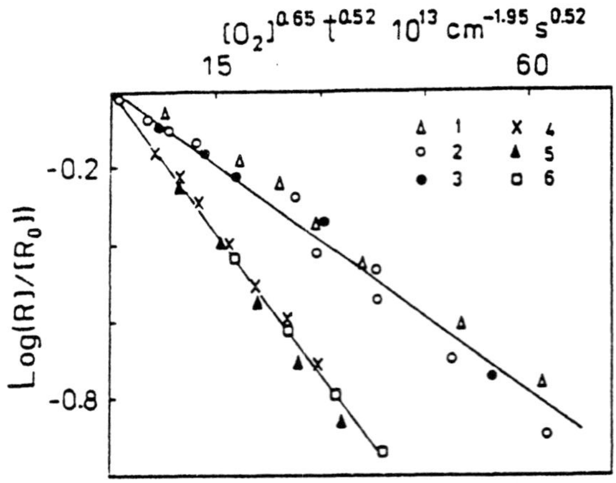

Figure 8. The kinetics of radical oxidation at T = 90K in anamorphoses testing the applicability of law (5). Oxygen concentration: 1—0.88×10^{18} cm^{-3}; 2—1.76×10^{18} cm^{-3}; 3—3.5×10^{18} cm^{-3}; 4—3.5×10^{19} cm^{-3}: 5—4.9×10^{19} cm^{-3}; 6—12.3×10^{19} cm^{-3}.

When oxygen concentration in our matrix is high enough, the total O_2 concentration in the sample can be much higher than the maximal gas concentration in traps. In this case the fact that a portion of gas does not participate in diffusion can be neglected. It might be expected that within the concentration range the oxidation kinetics for not annealed samples can also be described by law (5).

Figure 8 depicts the kinetics of radical oxidation in not annealed samples at T = 90K and $[O_2] = (0.88 - 220) \times 10^{19} \text{cm}^{-3}$ in anamorphoses testing the applicability of law (5). The figure demonstrates two applicability regions with oxygen concentrations $[O_2] < 3.5 \times 10^{18} \text{cm}^{-3}$ and $[O_2] > 3.5 \times 10^{19} \text{cm}^{-3}$. When $[O_2] = (0.35 - 3.5) \times 10^{19} \text{cm}^{-3}$ the kinetics does not obey law (5).

Thus the following can be assumed for the dissolved oxygen. At low concentrations $\left([O_2] < 3.5 \times 10^{18} \text{cm}^{-3}\right)$ the deep traps that prevent oxygen from participation in the diffusion process are filled to a small extent so that on the whole the amount of diffusion oxygen is proportional to the dissolved one. At large concentrations $\left([O_2] > 3.5 \times 10^{18} \text{cm}^{-3}\right)$ the deep traps are filled, their concentration being substantially lower than that of dissolved oxygen and the concentration of diffusion oxygen being proportional to the dissolved one. In the intermediate region the dependence of the diffusing oxygen on the dissolved one is nonlinear.

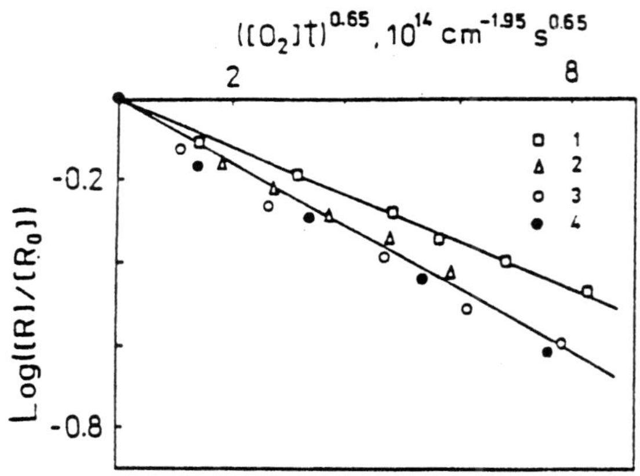

Figure 9. The kinetics of radical oxidation at 90K in the samples relaxed by destruction at T = 77K. Relaxation was performed before radical generation. Oxygen concentration: 1—0.35 × 10^{19} cm–3; 2—0.87 × 10^{19} cm–3; 3—1.3 × 10^{19} cm^{-3}; 4—3.5 x 10^{19} cm^{-3}.

The kinetic curves in this intermediate region in the samples subjected to destruction at 77K or annealed at 122K coincide in polychromatic coordinates. This is demonstrated in Figure 9 using the kinetics observed in the samples subjected to destruction as an example. The changes in the kinetic law due to either destruction of samples or annealing at 122K can be assigned to the relaxation of matrix structure. This relaxation destroys the traps and sets free the captured oxygen. At low concentrations the deviation from laws (3) is noticeable. This is denoted by curve 1 in Figure 9. The deviation is, probably, related to the fact that the destruction so as the high-temperature annealing for 2 hours, fails to cause a full relaxation of samples. Consequently, at 90K the additional relaxation processes are observed in the matrix that decrease the oxidation rate (Figure 2). On the other hand, this deviation can be caused by the fact that not all deep oxygen traps can be irreversibly destroyed by annealing at 122K or by destruction of matrix at 77K.

Figure 10 depicts the dependence of the relation between the concentration of free oxygen and the total oxygen concentration in the sample (γ) on the logarithm of the total oxygen concentration in matrix. This relation was calculated by three methods. In the first case

γ was calculated from the kinetics in the samples not subject to relaxation:

$$\frac{[R]}{[R_0]} = \left(-K_2\left(\gamma[O_2]\right)^{0.65} t^{0.52}\right) \tag{11}$$

It was assumed that $\gamma = 0.9$ at $[O_2] = 3.5 \times 10^{19}$ cm^{-3}. In the second case the relation between the concentration of trapped oxygen and the total concentration was calculated from the kinetic curves for oxidation reaction in the samples with $[O_2] = (1.32 - 13.2) \times 10^{19}$ cm^{-3} preliminarily annealed at reaction temperature for 20.5 hours (annealing time was 5.1 times longer than the maximum reaction time in the annealed samples) using equation (12):

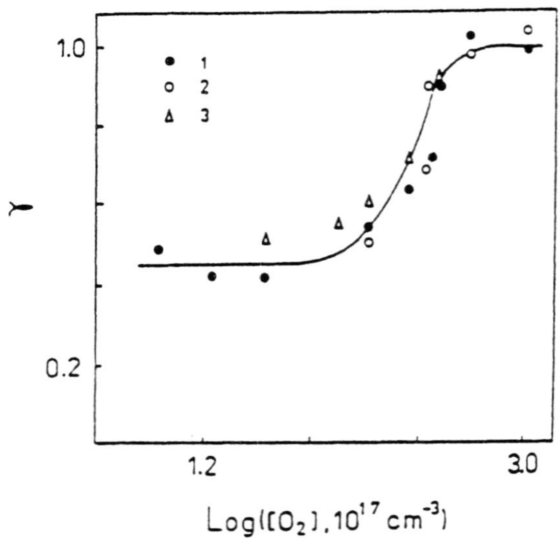

Figure 10. The dependence of the relation between the concentration of free oxygen and the total oxygen concentration in the sample on the logarithm of total oxygen concentration in matrix, calculated: 1—from the oxidation kinetics at T = 90K in not annealed samples; 2—from the oxidation kinetics at 90K in the samples preliminarily annealed at reaction temperature for 20.5 hours; 3—from the effect of oxidation rate increase at T = 90K due to preliminary annealing at T = 122K.

$$\frac{[R]}{[R_0]} = \exp\left(-K_1\left(\gamma[O_2]t\right)^{0.65}\right) \tag{12}$$

It was also assumed that $\gamma = 0.9$ at $[O_2] = 3.5 \times 10^{19}$ cm^{-3}. In the third case this relation was calculated from the effect of the oxidation rate increase in the sample subjected to destruction 77K or in the samples annealed at 122K before radical generation:

$$A\gamma^{0.65} = \frac{K'}{K''}$$

where K' is the slope of the kinetic curves in anamorphoses:

$$\log([R]/[R_0]) \div ([O_2]t)^{0.65}$$

in the samples annealed at 122K or subjected to mechanical destruction at 77K; K'' is the slopes of the kinetic curves in anamorphoses:

$$\log([R]/[R_0]) \div [O_2]^{0.65} t^{0.52}$$

in the samples not subjected to relaxation. Coefficient A was calculated from the normalization condition: $\gamma = 0.9$ at $[O_2] = 3.5 \times 10^{19}$ cm^{-3}.

As seen from Figure 10 there is a fair agreement between the relations of γ obtained using different methods. Thus, the data resulting from the above treatment of the kinetic curves are in satisfactory agreement with our hypothesis on the existence of a certain amount of deep oxygen traps in the matrix and on the destruction of these traps either by annealing at 122K or by mechanical destruction of matrix at 77K.

Consider now in more detail the influence of prolonged annealing at 90K on the oxidation kinetics at the same temperature. The reaction kinetics in the samples preliminarily annealed at T = 90K for 22 days is described by polychromatic law (12) with oxygen concentration in the matrix $[O_2] = (0.175 - 3.5) \times 10^{19}$ cm^{-3} (Figure 11) and γ varying as shown in Figure 10. Besides, it was found that the influence of the preliminary annealing at reaction temperature on the kinetic process at 90K obeys law (10) (Figure 12) where K_1 and K_2 were calculated using $\gamma\,[O_2]$ for $[O_2]$. Thus, description (5–10)holds for the reaction at both temperatures (98 and 90K).

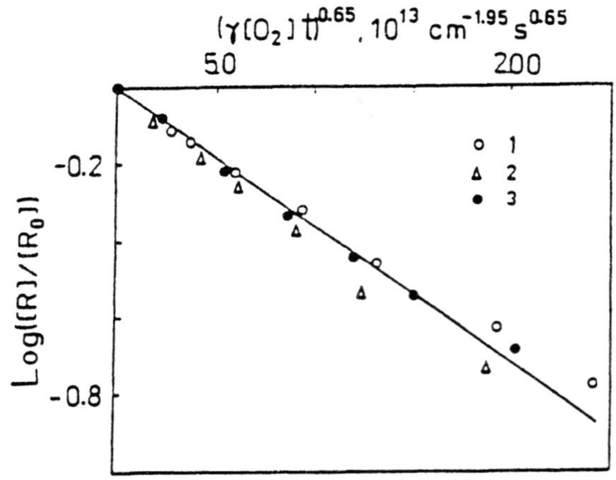

Figure 11. The kinetics of oxidation reaction at T = 90K in the samples preliminarily annealed at reaction temperature for 22 days in anamorphoses testing the applicability of polychromatic description. Oxygen concentration ($[O_2]$): 1—1.76 × 10^{18} cm^{-3}; 2—3.5 × 10^{18} cm^{-3}; 3—3.5 × 10^{19} cm^{-3}.

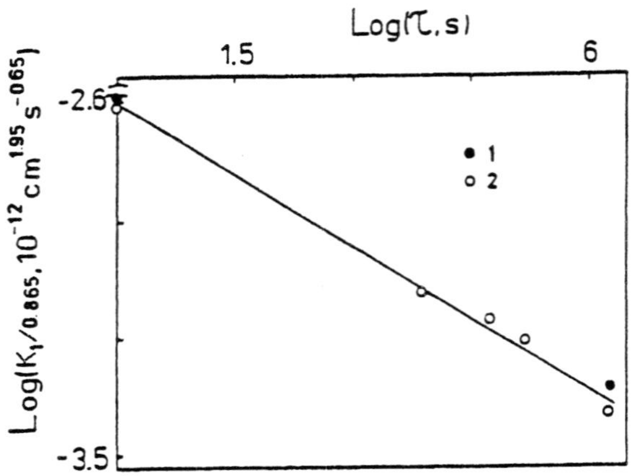

Figure 12. The dependence of log k_1 for the reaction at T = 90K on the logarithm of the preliminary annealing duration at reaction temperature. Oxygen concentration ($[O_2]$):1— 3.5 × 10^{18} cm^{-3}; 2—3.5 × 10^{19} cm^{-3}. Points on the ordinate—log k_2 (10^{12} $cm^{1.95}$ $s^{-0.52}$). The slope of straightline is 0.13.

The kinetics of radical oxidation in C_4D_9OD matrix has been studied in References [8–9]. The authors managed to describe the oxidation kinetics in not annealed samples and in the samples preliminarily annealed at reaction temperature using law (1) on the assumption that: (i) all matrix oxygen participates in diffusion; (ii) the constant depending on the reaction and annealing times is divided into two factors:

$$K(t,\tau) = K_\alpha(t+\tau)K_\beta(t)$$

where t is the reaction time; τ is the annealing duration at reaction temperature before radical generation. The above experiments testify this description to be unsatisfactory because not all matrix oxygen participates in diffusion. The error was revealed owing to a wider range of oxygen concentrations used in our work. Dependence $\tau^{0.13}$ was also determined again (according to [8] it is $\tau^{0.09}$).

Discuss now the possible reasons for rate constant decrease during annealing at 90 and 98K (7). As has been mentioned, two hypotheses can be discussed. In terms of the first one, the rate constant decrease observed during annealing is due to the matrix structure relaxation. Dependence (7) is in agreement with this hypothesis because in this case the slowing down of ki is determined by the time of annealing. The second hypothesis assumes that during annealing the oxygen molecules are redistributed over deep traps. In this case the matrix structure remains unchanged. The decrease in ki is determined by the number of jumps. Hence:

$$\frac{k_i(t)}{k_i(t+\tau)} = \frac{J_i(t)}{J_i(t+\tau)} = F_i(\eta_i(t,\tau))$$

where $J_i(t)$ is the frequency of jumps at moment t; $J_i(t+\tau)$ is the frequency of jumps at moment $t+\tau$; $\eta_i(t,\tau)$ is the number of jumps performed by molecule from the i-th group from moment t to $t+\tau$. It is evident that for $i \neq j$:

$$\eta_i(t,\tau) \neq \eta_j(t,\tau)$$

However, as follows from (7):

$$F_i(\eta_i(t,\tau)) = F_j(\eta_i(t,\tau)) \qquad (14)$$

for the different i and j. In addition, the values for $F_i(\eta_i(t,\tau))$ for the reaction at 90K are the same as those for a similar function, $F'_i(\eta_i(t,\tau))$

for the reaction at 98K although the number of jumps performed by the molecules of this group t to $t + \tau$ is different at different temperature.

Thus, in the framework of this model the reasons are assumed to exist that determine both (14) and the temperature-dependence of the $F_i(\eta_i(t,\tau))$ values. Therefore from our viewpoint as assumption that the slowing down of k during annealing is due to the structural matrix relaxation is most probable.

4. CONCLUSION

It was shown [27–30] that the glassy polymeric films contain deep traps where the dissolved in these films gas loses its ability to participate in diffusion. Experimental data have been interpreted in terms of this assumption. Thus the low-molecular alcohol glasses can also contain the deep traps where dissolved gas loses its ability to diffuse throughout the sample.

Our experimental data testify that in the n–C_4D_9OD matrix there are two temperature regions of thermal relaxation. In this case a high-temperature annealing (i.e. annealing at 122K) is related to the relaxation of matrix structure which causes the irreversible destruction of deep oxygen traps. Keeping the sample at T = 117K (i.e. in the intermediate temperature domain) leads to the losses of memory about the low-temperature annealing. Besides, it was found that the high-temperature annealing has the same influence on the oxidation kinetics as the mechanical destruction of the sample at 77K.

In our matrix the oxidation kinetics was described in terms of the polychromatic approach with taking into account the fact that within 90–100K the relaxation processes occur whose influence on the rate constant is described by the expression

$$k_i(t,\tau) = a_i(t,\tau)^{-0.2}$$

where k_i is the rate constant in the i-th group; t is the reaction time; t is the annealing duration at reaction temperature before radical generation. Thus, the relaxation processes occurring in the matrix during low-temperature annealing (i.e. annealing at T = (87–110K) slow down k_i. This can be related to both the redistribution of oxygen molecules during annealing over deep traps and the structural matrix relaxation. Using the above kinetic data we assume that the most probable reason for low-temperature annealing is the structural matrix relaxation.

In the framework of this hypothesis the data obtained testify to the fact that in the n-C_4D_9OD matrix there are two types of regions: with more porous and more dense packings with radicals being mainly in the regions with more porous packing.

REFERENCES

1. **Ya. S. Lebedev**, Kinetika i kataliz, Vol. 19, 1978, pp. 1367–1375 (Russian).
2. **N.M. Emanuel** and **A.L. Buchachenko**, "Chemical physics of Degradation and Stabilization of Polymers," VNU, Amsterdam, 1988, 368 p.
3. **V.A. Bagryansky** and **V.A. Tolkatchev**, Vysokomol. Soedineniya, Vol. B12, 1987, pp. 909–911 (Russia).
4. **V.M. Syutkin** and **V.A. Tolkatchev**, *React.* Kinet. Catal. Lett., Vol. 29, 1985, pp. 417–421.
5. **V.M. Zaskul'nikov** and **V.A. Tolkatchev**, Kinetika i kataliz., Vol. 20, 1979, p. 263 (Russian).
6. **V.M. Zaskul'nikov** and **V.A. Tolkatchev**, React. Kinet. Katal. Lett., Vol. 14, 1980, pp. 435–438.
7. **V.M. Zaskul'nikov, V.I. Vyazovkin, B.V. Bolshakov,** and **V.A. Tolkatchev**, Int. J. Chem. Kinet., Vol. 13, 1981, pp. 707–728.
8. **E. L. Zapadinskii, V.V. Korolev, N.M. Bazhin,** and **V.A. Tolkatchev**, React. Kinet. Catal. Lett., Vol. 37, 1988, pp. 477–479.
9. **V.M. Zaskul'nikov** and **V.A. Tolkatchev**, Kinetika i kataliz., Vol. 23, 1982, pp. 1063–1070 (Russian).
10. **V.A. Bagryansky, V.M. Zaskul'nikov,** and **V.A. Tolkatchev**, Chem. Phys., Vol. 78, 1983, pp. 41–48.
11. **V.I. Myagkikh, V.M. Zaskul'nikov,** and **V.A. Tolkatchev**, React. Kinet. Catal. Lett., Vol. 14, 1980, pp. 73–76.
12. **V.A. Bendersky, E.Ya. Misochko, A.A. Ovchinnikov,** and **P.G. Filippov**, Pis'ma v ZhTF., Vol. 32, 1980, pp. 429–431 (Russian).
13. **I.M. Barkalov, V.I. Goldanskii, D.P. Kiryukhin,** and **A.M. Zanin**, Int. Rev. Phys. Chem., Vol. 3, 1983, pp. 247–262.
14. **V.A. Bendersky, V.A. Titov,** and **P.G. Filippov**, Khimicheskaya fizika, Vol. 7, 1988, pp. 327–334 (Russian).
15. **V.A. Bendersky, E. Ya. Misochko, A.A. Ovchinnikov,** and P.G. **Filippov**, Zhurnal fizicheskoi khimii., Vol. 57, 1983, pp. 1079–1090 (Russian).
16. **V.A. Bendersky, E. Ya. Misochko, A.A. Ovchinnikov,** and **P.G. Filippov**, Khimicheskaya fizika., Vol. 5, 1982, pp. 685–691 (Russian).
17. **V.V. Barelko, I.N. Barkalov, D.A. Vaganov, A.M. Zanin,** and **D.P.Kiryukhin**, Dokl. AN SSSR., Vol. 264, 1982, pp. 99–102 (Russian).

18. **A.A. Ovchinnikov** and **V.A. Onischuk**, Khimicheskaya fizika, Vol. 3, 1984, pp. 511–526 (Russian).
19. **V.K. Ermolaev, Yu. N. Molin,** and **N. Ya. Buben**, Kinetika i kataliz., Vol. 3, 1962, pp. 314–321 (Russian).
20. **N. Scwarc**, J. chim. phys., Vol. 59, 1962, pp. 1067–1071.
21. **N. Scwarc**, J. chim. phys., Vol. 63, 1966, pp. 137–141.
22. **M.A. Neiss** and **J.E. Willard**, J. Phys. Chem., Vol. 79, 1975, pp. 783–794.
23. **V.M. Anisimov, I.R. Mardeleishvili,** and **N.I. Zaitsev**, Vysokomol. soedineniya, Vol. A30, 1988, pp. 1239–1243 (Russia).
24. "Handbook of solubility," Vol. 1, No. 1, Moskow, Izd. AN SSSR, 1962, 572 p. (Russian).
25. **S.V. Vasenkov, V.A. Bagryansky, V.V. Korolev,** and **V.A. Tolkatchev**, Int. J. Radiat. Phys. Chem., (to be published).
26. **V.A. Tolkatchev, I.I. Chkheidze,** and **N.Ya. Buben**, Dokl. AN SSSR, Vol. 147, 1962, pp. 643–644 (Russia).
27. **P. Meares**, J. Am. Chem. Soc., Vol. 76, 1954, pp. 3415–3422.
28. **R.M. Barrer, J.A. Barrie,** and **J. Slatet**, J. Polym. Sci., Vol. 27, 1958, pp. 177–197.
29. **T.M. Aminabhavi, U.S. Aithal,** and **S.S. Skukla**, J. Macromol. Sci., Vol. C28, 1988, pp. 421–427.
30. **W.J. Koros** and **D.R. Paul**, J. Polym. Sci., Vol. 16, 1978, pp. 1947–1963.
31. **S.A. Dzuba** and **Yu. D. Tsvetkov**, Zhurnal strukturnoi khimii, Vol. 28, 1987, pp. 15–38 (Russian).
32. **G.P. Johari**, J. Chem. Phys., Vol. 82, 1985, pp. 283–291.

KINETICS OF SOME HETEROGENEOUS CATALYTIC REACTIONS INVOLVING MOLECULAR HYDROGEN

N.V. Pavlenko, and A. I. Tripolskiĭ
L.V. Pisarzhevskii Institute of Physical Chemistry,
Ukrainian Academy of Sciences, Kiev

Abstract—The data are systematized on the kinetic patterns of the processes of vapor-phase hydrogenation of aliphatic carbonyl and nitrile compounds, as well as carbon mono-and dioxide, on transition metals. Various kinetic models of these reactions and their correspondence with the observed kinetic dependences are considered. The patterns of partial and exhaustive hydrogenation of the compounds studied are discussed.

1. INTRODUCTION

Heterogeneous catalytic processes involving molecular hydrogen include a great number of reactions that are of practical importance, especially if they can be carried out at moderate temperatures and atmospheric pressure.

Despite the vast body of works devoted to the addition of hydrogen to compounds of various classes, the kinetics of such reactions are not sufficiently studied, although these processes share a number of features that are essential for elucidating the mechanism of the hydrogenation processes, for developing and improving the relevant technologies. Another important problem remaining unsolved in the theory of technological processes is the dependence of the rate and the mechanism of catalytic hydrogenation on the structure and properties of the reacting molecules and the catalyst.

This review will mainly consider the works of recent years concerning the kinetic patterns of gas-phase hydrogenation of various compounds—aliphatic carbonyl and nitrile compounds, carbon mono and dioxide—in the presence of solid and deposited transition metals. As these compounds belong to different classes, one can hope that in the future this will help to reveal and formulate the essential common patterns in the kinetics of catalytic addition of molecular hydrogen.

Considering the kinetics of catalytic hydrogenation, account should be taken of some common peculiarities of these reactions listed below.

1. The hydrogenation reactions are, as a rule, exothermic. Therefore great care should be taken in kinetic studies to avoid temperature gradients.
2. These reactions are usually complicated multistage processes developing in various directions (via several independent routes). Therefore the kinetic data should be processed and interpreted using the concepts of the theory of complex reactions [1].
3. The design of kinetic experiments and data processing should take into account the nonsteady-state transitions caused by the influence of the reaction system on the catalyst. Therefore various kinetic comparisons should only be made upon attaining one and the same steady state of the catalyst surface.
4. The hydrogenation reactions involve factors giving rise to various forms of surface intermediates. The influence of these factors should be accounted for in interpreting the kinetic data.
5. Most of the hydrogenation reactions are reversible, therefore the kinetic studies must take into account the approach to equilibrium and the effect of the reverse reaction; as a rule, the kinetic analysis is preceded by thermodynamic calculations.

2. Thermodynamics of Some Hydrogenation Reactions

Usually the hydrogenation reactions are reversible and exothermic. Consequently, at high temperatures the reverse process—dehydrogenation—will dominate. Depending on the reaction conditions and the catalysts used, hydrogenation of carbonyl compounds can give rise to alcohols and hydrocarbons.

Reduction of acetone and propionic aldehyde in the presence of metal catalysts mainly yields the corresponding alcohol and propane:

$$C_3H_6O + H_2 = C_3H_8O \qquad (I)$$

$$C_3H_6O + 2H_2 = C_3H_8 + H_2O \tag{II}$$

Reduction of nitriles, in particular acetonitrile and propionitrile, yields primary (A_1), secondary (A_2), and tertiary (A_3) amines:

$$CH_3CN + 2H_2 = CH_3CH_2NH_2 \tag{III}$$

$$2CH_3CN + 4H_2 = (CH_3CH_2)_2NH + NH_3 \tag{IV}$$

$$3CH_3CN + 6H_2 = (CH_3CH_2)_3\,N + 2NH_3 \tag{V}$$

$$CH_3CH_2CN + 2H_2 = CH_3(CH_2)_2NH_2 \tag{VI}$$

The temperature dependence of the equilibrium constants calculated [2–4] from the standard free energy changes in reactions (I) to (VI) at 300–500K and 0.1 MPa is shown in Figure 1. It can be seen that in the temperature range studied all the above processes are thermodynamically favorable ($\log K_p > -1$). With increasing temperature the K_p values drop. The greater is the number of carbon atoms in the nitrile molecule the lower is the equilibrium constant of nitrile hydrogenation.

Complete reduction of carbonyl compounds (process II), unlike mild hydrogenation in reaction (I), is practically irreversible, as well as formation of tertiary amine is irreversible as compared with that of primary (III) and secondary (IV) amines.

In reactions leading to methane and its homologs:

$$CO + 3H_2 = CH_4 + H_2O \tag{VII}$$

$$nCO + (2n + 1)H_2 = C_nH_{2n+2} + nH_2O \ (n \geq 2) \tag{VIII}$$

$$CO_2 + 4H_2 = CH_4 + 2H_2O \tag{IX}$$

$$nCO_2 + (3n + 1)H_2 = C_nH_{2n+2} + nH_2O \ (n \geq 2) \tag{X}$$

the equilibrium constants decline with increasing temperature [5,6]; however, for all hydrocarbons they remain high enough up to about 700K. In contrast to processes (I)–(X), the equilibrium constant of partial reduction of carbon dioxide to carbon monoxide

$$CO_2 + H_2 = CO + H_2O \tag{XI}$$

is close to zero; with increasing temperature the thermodynamic probability of CO formation increases [6].

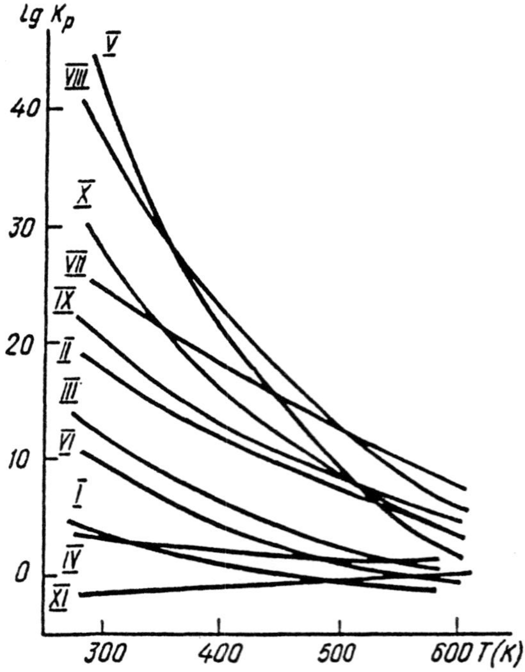

Figure 1. Temperature dependences of logK_P of reactions (I) to (XI). The K_P values for reactions (VIII) and (X) have been obtained for n = 2.

3. Kinetics of Hydrogenation of Aliphatic Carbonyl Compounds

For reaction (I) in a static system with films of pulverized transition metals it has been shown [7] that its rate (r) is independent of acetone concentration (P_A), depends in a complex manner on partial hydrogen tension $\left(P_{H_2}\right)$, and is decreased by the reaction product, isopropyl alcohol (P_P).For the reaction on Raney nickel, the following kinetic equation has been proposed [8]:

$$r = k \cdot P_A^{1.2} \cdot P_{H_2}^{1.5}$$

where k is the reaction rate constant. Another rate equation for reaction (I) on Raney nickel:

$$r = k \cdot P_A \sqrt{P_{H_2}} / \sqrt{P_P}$$

has been found in the work [9] where its kinetics were studied in the liquid phase (n-octane) and the influence of solvent was disregarded.

Reaction (I) has also been studied [10,11] on complex nickel catalysts ($Ni/ZnO \cdot Cr_2O_3$ and Ni/Al_2O_3), but as the conditions of their preparation are not given it is not clear whether the results obtained pertain to metallic Ni. These studies [10,11] were carried out under conditions that did not ensure catalyst isothermicity, and at high temperatures when the reverse reaction of isopropanol dehydrogenation became significant but was disregarded. It may be for this reason that the activation energy of the process became negative with increasing temperature.

The kinetic orders of reaction (I) have been determined [12,13], and exponential kinetic equations have been derived [13] to describe the process on a nonuniform surface in the region of reversibility.

A detailed study of the kinetics of vapor-phase hydrogenation of acetone on solid Ni, Co, Cu, and their alloys has been performed by Babkova et al. [14–17]. Kinetic measurements conducted in the presence of Ni black in the temperature range 323–363°K have shown [14] that the reaction rate appreciably increases with acetone partial tension, is less dependent on hydrogen concentration, and is negatively affected by isopropanol. The reaction orders in respect of its components are variable and depend on component concentrations and temperature.

Babkova et al. [14] have suggested that the reaction involved hydrogen atoms adsorbed on Ni in proton form, and the rate of the process is limited by the surface interaction of adsorbed acetone molecules and hydrogen atoms. In respect of adsorption of both reagents the metal surface is assumed to be effectively uniform. The nonplanary acetone molecules do not fill all the adsorption sites of the metal surface even at complete saturation. It is suggested therefore that hydrogen atoms are adsorbed on the surface in the spaces between acetone molecules. Acetone and hydrogen are believed to be adsorbed on different active sites of the catalyst. On the basis of kinetic data an adsorption mechanism of the reaction is proposed, and a corresponding kinetic equation is derived which provides a good description of the experimental data [14].

The kinetic dependences and the mechanism of reaction (I) typical of nickel are maintained with a cobalt catalyst and nickel-cobalt alloys [15], but the reaction rate and the rate constant decline with increasing atomic portion of cobalt in Ni-Co alloys (over 50%).

For a carrier-free copper catalyst, the same authors [16,17] have found that the reaction rate is practically independent of P_A, increases with P_{H_2}, and is decreased by the reaction product, isopropyl alcohol. The reaction order in respect of isopropanol depends on the alcohol concentration in the reaction mixture and on temperature.

Basing on the data [18,19] that molecular hydrogen is not adsorbed on copper below 573°K, Babkova et al. [16] have considered different versions of an impact mechanism of acetone hydrogenation. According to this mechanism, the limiting step of the process is the interaction of a hydrogen molecule from the gas phase with an adsorbed ketone molecule.

Quite interesting is the suggestion [16] that acetone hydrogenation on copper can be described by the mechanism put forward by Twigg [20] for ethylene hydrogenation on nickel and applied by Apelbaum and Temkin [21] to ethylene hydrogenation on a palladium film. This mechanism implies that hydrogen from the gas phase interacts with an adsorbed acetone molecule to yield a semihydrogenated form of acetone and a hydrogen atom adsorbed on a free site of the Cu surface. The ability of copper to adsorb atomic hydrogen has been reliably established [22]. The limiting step of the reaction is the interaction of the adsorbed hydrogen atom with the semihydrogenated acetone. On the basis of this mechanism the authors [16] have proposed the following kinetic equation describing the experimental data:

$$r = k_4 \frac{K_3 b_A P_A \sqrt{b_{H_2} P_{H_2}}}{1 + b_A P_A + K_3 b_A P_A \sqrt{b_{H_2} P_{H_2}} + b_P P_P} \cdot \frac{\sqrt{b_{H_2} P_{H_2}}}{1 + \sqrt{b_{H_2} P_{H_2}}}$$

where K_3 is the equilibrium constant of the third step, and b_i are the adsorption coefficients.

In this work [16], unlike the previous one [14], the authors suggest that the limiting step of acetone hydrogenation is the same on Ni and Cu; yet the publication [16] does not provide convincing arguments in favor of that the adsorbed acetone molecule is more reactive than its semihydrogenated form, and that addition of the second hydrogen atom is the slowest step.

The similarity of the kinetic patterns of reaction (I) catalyzed by nickel and by nickel–copper alloys with a Cu content up to some 60 at.% allows a suggestion [17] that the scheme of the reaction mechanism on a Ni catalyst [16] is also true for alloyed catalysts of the above composition. At copper concentrations in the alloy over 60% the process can be described by the Twigg mechanism proposed for the 100%-copper catalyst [16].

Studies on the kinetics of reversible acetone hydrogenation on a nickel catalyst [13,23] reveal that the reaction proceeds via two routes: the adsorbed ketone molecule interacts with a hydrogen atom, whereas the semihydrogenated form reacts either with hydrogen in the gas phase or with the second hydrogen atom. Dehydrogenation of isopropyl alcohol includes interaction of isopropanol from the gas phase with dissociatively adsorbed hydrogen. The corresponding kinetic equation looks as follows:

$$r = k\frac{P_A\left(1+k\sqrt{P_{H_2}}\right)\cdot\beta}{1+k_1P_A+K_2P_A\sqrt{P_{H_2}}+k_3\sqrt{P_{H_2}}}$$

where k_i are the constants and β is the correction for the influence of the reverse reaction.

The kinetics of acetone hydrogenation has been studied in a differential flow reactor at 423°K in the presence of Cu, Pt, Pd, and Rh applied on kieselguhr [24]. Computer analysis has been done for 29 kinetic equations corresponding to the Langmuir-Hinshelwood mechanism. The experimental dependences obtained in the presence of the copper catalyst are described by a kinetic model [24]

$$r = k\frac{b_AP_Ab_{H_2}P_{H_2}}{\left(1+\sqrt{b_AP_A}+\sqrt{b_{H_2}P_{H_2}}\right)^4}$$

implying that molecular adsorbed hydrogen reacts with dissociatively adsorbed acetone molecules. Regretfully, the authors [24] do not specify what is meant by dissociative adsorption of ketone on the surface of copper and what the mechanism of this process is.

The kinetic patterns obtained with metals of the platinum group (Pt, Rh, Pd) are in compliance with the equations based on the concepts of interaction between molecular adsorbed acetone and adsorbed hydrogen:

$$r = k\frac{b_AP_Ab_{H_2}P_{H_2}}{\left(1+b_AP_A+b_{H_2}P_{H_2}\right)^4}$$

$$r = k\frac{b_AP_Ab_{H_2}P_{H_2}}{\left(1+b_AP_A+\sqrt{b_{H_2}P_{H_2}}\right)^4}$$

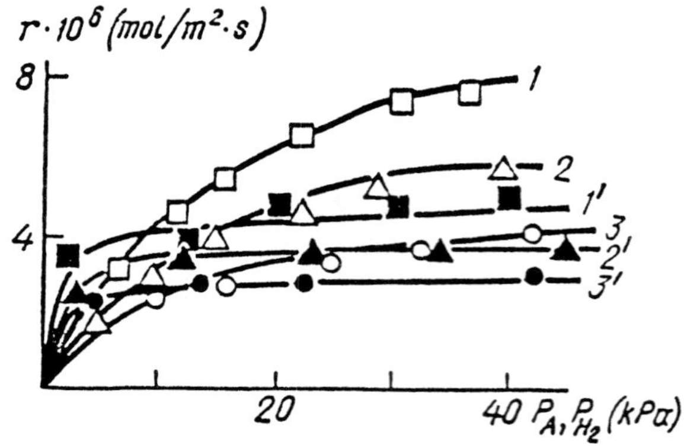

Figure 2. Dependences of the rate of reaction (I) on Pt/Al_2O_3 on (1, 2, 3) P_A and (1', 2', 3') P_{H_2} at (1,1') 365°K, (2, 2') 348°K, and (3,3') 337°K.

It has been assumed [24] that on the surface of the catalysts mentioned, hydrogen is adsorbed in the molecular form. This contradicts numerous data in the literature that the atomic form of hydrogen adsorption is predominant on the surface of metals of the platinum group. Besides, at temperatures above 400°K it is necessary to take into account the reversibility of acetone hydrogenation, which has been neglected in the work of Simonicova et al. [24].

We have carried out systematic studies on vapor-phase hydrogenation of acetone on applied and solid transition metals: Fe, Co, Ni, Ru, Rh, Pd, Ir, Pt, and Cu [25–33]. A detailed analysis of the kinetics of reaction (I) at different temperatures reveals that in all cases with increasing reagent concentrations the rate of the process increases, reaching a plateau. Figure 2 depicts the typical kinetic dependences obtained in the presence of catalyst Pt/Al_2O_3. The reaction product does not hinder the process on any catalysts except the copper and ruthenium ones, where the rate declines with increasing P_P [26,27].

Analysis of the kinetic and adsorption data has allowed us to envisage the following scheme for the mechanism of reaction (I) [25,27]:

$$\begin{aligned}
&(1)\ H_2 + 2Z_H = 2Z_H H \\
&(2)\ (CH_3)_2CO + Z_A = Z_A(CH_3)_2CO \\
&(3)\ Z_A(CH_3)_2CO + Z_H H \rightarrow Z_A(CH_3)_2COH + Z_H \\
&(4(\ Z_A(CH_3)_2COH + Z_H H \rightarrow (CH_3)_2CHOH + Z_H + Z_A
\end{aligned} \qquad \text{(XII)}$$

where Z_A and Z_H are the surface adsorption sites for acetone and hydrogen, respectively. The first two steps reach an equilibrium, and the limiting step is the interaction of the adsorbed organic compound with $Z_H H$. Here two variants are possible. If addition of the first hydrogen atom to adsorbed acetone is limiting, the kinetic equation is

$$r = k_3 \frac{b_A P_A}{1 + b_A P_A} \cdot \frac{\sqrt{b_{H_2} P_{H_2}}}{1 + \sqrt{b_{H_2} P_{H_2}}} \tag{1}$$

where k_3 is the rate constant of the third step. If the last step (addition of hydrogen to the adsorbed "semihydrogenated" form of acetone) is limiting, while step (3) reaches an equilibrium, the kinetic equation acquires the following form:

$$r = k_4 \frac{K_3 b_A P_A \sqrt{b_{H_2} P_{H_2}}}{1 + b_A P_A + K_3 b_A P_A \sqrt{b_{H_2} P_{H_2}}} \cdot \frac{\sqrt{b_{H_2} P_{H_2}}}{1 + \sqrt{b_{H_2} P_{H_2}}} \tag{2}$$

where k_4 is the rate constant of the limiting (last) step and K_3 is the equilibrium constant of the third step.

Quantitative comparisons demonstrate that equation (1), unlike equation (2), is in good fit to the experimental data for all catalysts we have tested.

A study of the kinetics of the direct (I) and the reverse reaction (isopropanol dehydrogenation):

$$(CH_3)_2CHOH = (CH_3)_2CO + H_2 \tag{XIII}$$

on Pt/Al_2O_3 as a catalyst [32] has made it possible to detail the mechanism (XII). For reaction (XIII) it has been found that with increasing P_P at moderate P_{H_2} the reaction rate increases to a plateau (Figure 3). Under an excess of hydrogen the reaction order in respect of alcohol increases, approximating the first order. The products—acetone and hydrogen—are inhibitors for process (XIII) (Figure 4).

On the strength of the whole body of data obtained, the following mechanism of reversible addition of hydrogen to the carbonyl group has been put forward [32,33]:

$$(1)\ H_2 + 2Z_H = 2Z_HH$$
$$(2)\ (CH_3)_2CO + Z_A = Z_A(CH_3)_2CO$$
$$(3)\ Z_A(CH_3)_2CO + Z_HH \rightleftarrows Z_A(CH_3)_2COH + Z_H$$
$$(4)\ Z_A(CH_3)_2COH + Z_HH \rightleftarrows Z_H(CH_3)_2CHOH + Z_A$$
$$(5)\ Z_H(CH_3)_2CHOH = (CH_3)_2CHOH + Z_H \qquad \text{(XIV)}$$

In this mechanism, steps (3) and (4) are reversible but not equilibrium ones. Step (5) reaches an equilibrium, with isopropanol molecules being adsorbed on the same sites Z_H as hydrogen.

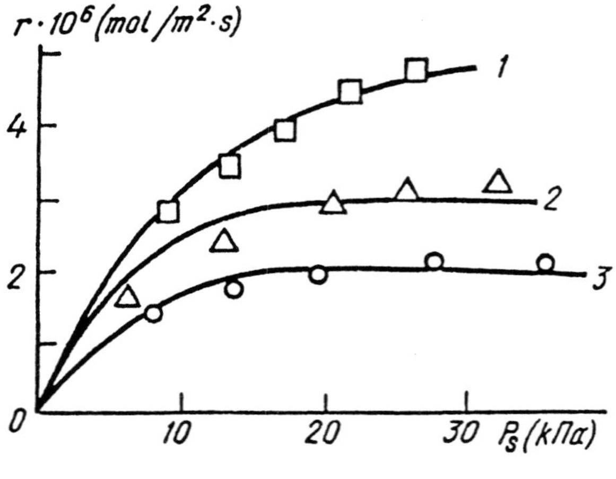

Figure 3. Dependences between Pp and the rate of reaction (XIII) on Pt/Al_2O_3 at (1) 423°K, (2) 404°K, and (3) 387°K.

The observed rate of acetone conversion under steady-state conditions is:

$$r = r_3 - r_{-3} = r_4 = r_{-4} \qquad (3)$$

where $r_{\pm i}$ are the rates of the steps in the forward and the reverse directions. Hence, applying the law of acting surfaces to r_i, r_{-i}, we obtain

$$k_3\Theta_A\Theta_H - k_{-3}\Theta_{AH}\Theta_H^0 = k_4\Theta_{AH}\Theta_H - k_{-4}\Theta_P\Theta_A^0 \qquad (4)$$

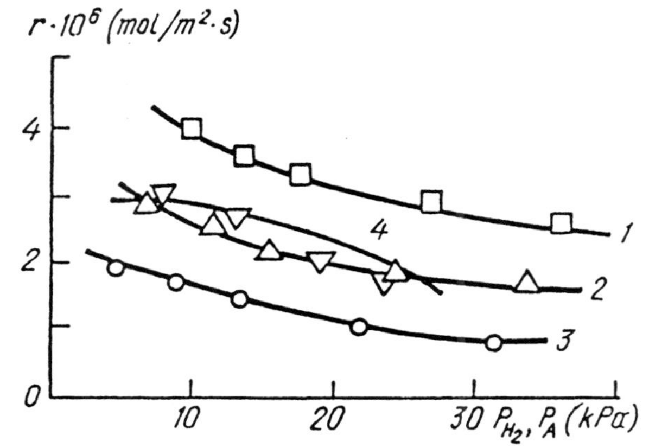

Figure 4. Dependences of the rate of reaction (XIII) on Pt/Al_2O_3 on (1, 2, 3,) P_{H_2} and (4) P_A at (1) 423°K, (2, 4) 404°K, and (3) 387°K.

where Θ_A is the occupancy of the Z_A sites by acetone, Θ_{AH} is the same for semihydrogenated acetone, Θ_A^o is the portion of free Z_A sites; Θ_H is the occupancy of the Z_H sites by hydrogen and Θ_p by the product (alcohol), is the portion of free Z_H sites; $k_{\pm i}$ are the rate constants for the corresponding steps. From equation (4) it follows that

$$\Theta_{AH} = \frac{k_3\Theta_A\Theta_H + k_{-4}\Theta_P\Theta_A^o}{k_{-3}\Theta_H^o + k_4\Theta_H} \qquad (5)$$

The equilibrium conditions of steps (1), (2), and (5) yield

$$\Theta_H = \Theta_H^o\sqrt{b_{H_2}P_{H_2}} \qquad (6)$$

$$\Theta_P = \Theta_H^o b_P P_P \qquad (7)$$

$$\Theta_A = \Theta_A^o b_A P_A \qquad (8)$$

Using the material balance equation

$$\Theta_H^o + \Theta_H^o\sqrt{b_{H_2}P_{H_2}} + \Theta_H^o b_P P_P = 1$$

we obtain

$$\Theta_H^o = \left(1 + \sqrt{b_{H_2} P_{H_2}} + b_P P_P\right)^{-1} \tag{9}$$

Substituting expressions (6) to (8) into equation (5) and taking account of equation (9), we arrive at

$$\Theta_{AH} = \Theta_A^o \cdot \Psi \tag{10}$$

where

$$\Psi = \frac{k_3 b_A P_A + k_{-4} b_P P_P}{k_{-3} + k_4 \sqrt{b_{H_2} P_{H_2}}} \tag{11}$$

The value Ψ characterizes the extent of reversibility of the reaction: if steps (3) and (4) are irreversible and step (3) is the limiting one ($k_3 << k_4$), then $\Psi \to 0$, which corresponds to mechanism (XII).

The material balance equation for sites Z_A

$$\Theta_A + \Theta_{AH} + \Theta_A^o = 1$$

together with equations (8) and (9) gives

$$\Theta_A^o = 1/(1 + b_A P_A + \Psi) \tag{12}$$

whence

$$\Theta_A = b_A P_A/(1 + b_A P_A + \Psi)$$

$$\Theta_{AH} = \Psi/(1 + b_A P_A + \Psi)$$

$$\Theta_H = \sqrt{b_{H_2} P_{H_2}}\Big/\left(1 + \sqrt{b_{H_2} P_{H_2}} + b_P P_P\right)$$

$$\Theta_P = b_P P_P\Big/\left(1 + \sqrt{b_{H_2} P_{H_2}} + b_P P_P\right)$$

Then the kinetic equation of reversible acetone hydrogenation corresponding to mechanism (XIV) acquires the form

$$r = r_3 - r_{-3} = k_3 \frac{b_A P_A}{1 + b_A P_A + \Psi} \cdot \frac{\sqrt{b_{H_2} P_{H_2}}}{1 + \sqrt{b_{H_2} P_{H_2}} + b_P P_P} - k_{-3} \frac{\Psi}{1 + b_A P_A + \Psi} \cdot \frac{1}{1 + \sqrt{b_{H_2} P_{H_2}} + b_P P_P} \tag{13}$$

It is only natural to regard equation (13) as the difference between the rates of reaction (I) in the forward and the reverse directions; i.e., the rate of acetone hydrogenation is equal to

$$\vec{r} = k_3 \frac{b_A P_A}{1 + b_A P_A + \Psi} \cdot \frac{\sqrt{b_{H_2} P_{H_2}}}{1 + \sqrt{b_{H_2} P_{H_2}} + b_P P_P} \tag{14}$$

while the rate of isopropanol dehydrogenation is

$$\overleftarrow{r} = k_{-3} \frac{\Psi}{1 + b_A P_A + \Psi} \cdot \frac{1}{1 + \sqrt{b_{H_2} P_{H_2}} + b_P P_P} \tag{15}$$

The kinetic model obtained satisfies the principle of detailed equilibrium. At equilibrium we have $\vec{r} = \overleftarrow{r}$, so on the basis of equations (14) and (15) we obtain

$$k_3 b_A P_A \sqrt{b_{H_2} P_{H_2}} = k_{-3} \cdot \Psi$$

and

$$\Psi = K_3 b_A P_A \sqrt{b_{H_2} P_{H_2}} \tag{16}$$

where $K_3 = k_3/k_{-3}$ is the equilibrium constant of the third step. On the other hand,

$$K_3 = \frac{\Theta_{AH} \cdot \Theta_H^o}{\Theta_A \cdot \Theta_H} = \Psi \Big/ b_A P_A \sqrt{b_{H_2} P_{H_2}} \tag{17}$$

and substitution of (17) into (16) gives an identity.

At low temperatures, when reaction (I) is practically irreversible ($\psi \to 0$), and at elevated hydrogen concentrations impeding alcohol adsorption

$$\left(\sqrt{b_{H_2} P_{H_2}} >> b_P P_P\right),$$

equation (13) degenerates into the kinetic equation (1) which nicely fits the experimental data [25, 27–33] on acetone hydrogenation on transition metals except Cu and Ru. With the latter catalysts, account should be taken of product inhibition of reaction (I) by isopropanol, and also of the fact that the process involves the molecular form of adsorbed hydrogen which we have identified by the programmed thermodesorption technique [27,34]. Therefore mechanism (XIV) changes somewhat:

$$
\begin{aligned}
&(1)\ H_2 + Z_H = Z_H H_2 \\
&(2)\ (CH_3)_2CO + Z_A = Z_A(CH_3)_2CO \\
&(3)\ Z_A(CH_3)_2CO + Z_H H_2 \overset{\rightarrow}{\underset{\leftarrow}{}} Z_H(CH_3)_2CHOH + Z_A\ . \\
&(4)\ Z_H(CH_3)_2CHOH = (CH_3)_2CHOH + Z_H
\end{aligned}
\qquad (XV)
$$

Here the limiting step is also the interaction of adsorbed initial substances; other steps reach equilibrium. The kinetic equation corresponding to mechanism (XV)

$$r = k_3 \frac{b_A P_A}{1 + b_A P_A} \cdot \frac{b_{H_2} P_{H_2}}{1 + b_{H_2} P_{H_2} + b_P P_P} \qquad (18)$$

is in good correspondence with the experimental data for Cu and Ru catalysts [26,27].

During acetone hydrogenation in the temperature range 300 to 370°K on deposited metals, formation of propane was observed [27] in the initial period of the reaction only with Pt and Ir. To explain the selectivity with respect to propane, it is sufficient to supplement the scheme (XIV) with a complex step of further interaction of adsorbed alcohol with adsorbed atoms:

$$(6)\ Z_H(CH_3)_2CHOH + Z_H H \rightarrow (Int) \xrightarrow[\text{rapid}]{Z_H H} C_3H_8 + H_2O + 3Z_H$$

where (Int) is an intermediate. Then the selectivity for hydrocarbon is

$$S_{C_3H_8} = k_6\Theta_H / (k_6\Theta_H + k_5)$$

At high enough P_{H_2} when $\Theta_H \approx 1$, we obtain

$$S_{C_3H_8} = \left[1 + (k_5 k_6)\right]^{-1}$$

Examination of individual steps of the mechanism of acetone hydrogenation on transition metals has shown [33] that for complete reduction of the carbonyl group of a ketone, the latter should be adsorbed in the form of a donor–acceptor complex, where the molecule is connected with the catalyst via the unshared electron pair of the carbonyl oxygen. Bearing in mind that acetone and other carbonyl compounds are weak Lewis bases, we have achieved the necessary form of adsorption and have reversed the direction of the reaction by using a bifunctional catalyst that, besides the hydrogen-activating transition metal, contains strongly acidic Lewis sites Z_L to which the carbonyl oxygen binds at the expense of the unshared electron pair. We have proposed [35] the following two-route mechanism of complete reduction of the acetone carbonyl group (reaction II) on a bifunctional catalyst:

$$
\begin{aligned}
&(1)\ H_2 + 2Z_H = 2Z_H H \\
&(2)\ (CH_3)_2CO + Z_L = Z_L(CH_3)_2CO \\
&(3)\ Z_L(CH_3)_2CO + Z_H H = Z_L(CH_3)_2COH + Z_H \\
&(4)\ Z_L(CH_3)_2COH + Z_H H = Z_L(CH_3)_2CHOH + Z_H \\
&(5)\ Z_L(CH_3)_2CHOH = (CH_3)_2CHOH + Z_L \\
&(6)\ Z_L(CH_3)COH + Z_H H \rightarrow Z_L C_3H_6 + H_2O + Z_H \\
&(7)\ Z_L C_3H_6 + Z_H H \rightarrow Z_L C_3H_7 \xrightarrow{Z_H H} C_3H_8 + Z_L + 2Z_H
\end{aligned}
\qquad \text{(XVI)}
$$

Steps (1) to (5) giving rise to isopropanol reach an equilibrium. Step (3) yields an intermediate $Z_L(CH_3)_2COH$ common for the syntheses of alcohol and hydrocarbon. The total process (II) is limited by step (6) in which addition of the hydrogen atom to the adsorbed semihydrogenated form of acetone is accompanied by dehydration and formation of adsorbed propylene ($Z_LC_3H_6$) in the form of a π-allyl complex [36]. Further, the adsorbed olefin is rapidly hydrogenated to propane.

The kinetic equation corresponding to mechanism (XVI)

$$
r = k_3 \frac{K_3 b_A P_A \sqrt{b_{H_2} P_{H_2}}}{1 + b_A P_A \cdot \left(1 + K_3 \sqrt{b_{H_2} P_{H_2}}\right) + b_P P_P} \cdot \frac{\sqrt{b_{H_2} P_{H_2}}}{1 + \sqrt{b_{H_2} P_{H_2}}} \qquad (19)
$$

describes well the experimental data [35] on the kinetics of reaction (II).

The data obtained have been used to calculate the kinetic and thermodynamic characteristics of the steps of mechanisms (XII) and (XIV) to (XVI): rate constants, adsorption coefficients, activation energies, adsorption heats and enthropies, and the pre-exponential factors in the rate constants [33]. Analysis of these values provides new arguments in favor of the proposed mechanisms. For example, as the heat effect of reaction (I) is equal to the sum of the heat effects of individual steps ($Q = \Sigma q_i$ where $q_i = E_{-i} - E_i$), the value calculated from the kinetic data (Q ≈ 30 kJ/mole) is close to that calculated for reaction (I) from the heats of acetone and isopropanol formation [5]. Similarly, the values of the equilibrium constant $K\Sigma$ of reaction (I) ($K\Sigma$ = $b_A b_{H_2} K_3 K_4 b_P^{-1}$ where $K_3 = k_3/k_{-3}$, $K_4 = k_4/K^{-4}$) found from the kinetic data agree with those obtained by direct thermodynamic calculations [2].

Analysis of the values of the pre-exponential factors of the rate constants of nonequilibrium steps (3) and (4) using the transition state theory demonstrates that the portion of active catalytic sites is much lower than unity. This is in line with the idea that formation of activated complexes in these steps requires a combination of simple sites Z_H with complex sites Z_A whose portion is low.

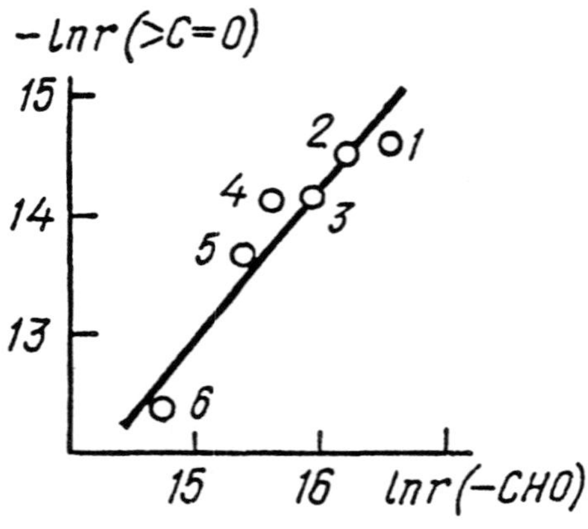

Figure 5. Comparison of the specific catalytic activity of metals in hydrogenation of acetone (364°K) and propionic aldehyde (405°K): (1) Co, (2) Cu, (3) Ni, (4) Rh, (5) Ir, (6) Pt.

Comparison of the kinetics of hydrogenation of ketones and aldehydes with the same number of carbon atoms on the same catalysts has revealed the following [37]. In both cases reaction (II) proceeds by mechanism (XIV). The process is limited by the addition of the first hydrogen atom to the carbonyl group of the adsorbed molecule; ketone and aldehyde conversion rates are described by the same equation (13). In hydrogenation of propanal $\left(405^\circ K, P_R = 20, P_{H_2} = 50 kPa\right)$ the transition metals display the following series of activity: Pt > Ir > Rh > Ni

> Cu > Co, which is identical to that observed in acetone hydrogenation (Figure 5). It must be mentioned that Subbotin et al. [38–40] substantiate another point of view, according to which hydrogenation of n-butyric aldehyde is limited by the addition of the second hydrogen atom.

In the presence of a smelted iron catalyst, the kinetics of hydrogenation of aliphatic and alicyclic ketones has been described by an exponential equation [41]:

$$r = k \cdot P_{\text{ket}} \cdot \sqrt{P_{H_2}} \tag{20}$$

On the basis of kinetic data on octanone-2 hydrogenation the following scheme has been proposed [42], implying reagent adsorption on the same sites Z of the Fe catalyst:

$$\begin{aligned}
&(1)\ H_2 + 2Z = 2ZH \\
&(2)\ R_1COR_2 + Z = ZR_1COR_2 \\
&(3)\ ZR_1COR_2 + ZH \rightarrow ZR_1COHR_2 + Z \\
&(4)\ ZR_1COHR_2 + ZH = ZR_1CHOHR_2 + Z \\
&(5)\ ZR_1CHOHR_2 = R_1CHOHR_2 + Z
\end{aligned} \tag{XVII}$$

Equation (20) has been used [41] to optimize the hydrogenation process for alicyclic ketones (C_5, C_6) formed in the synthesis of alcohols by oxidation of higher paraffins.

The mechanism of hydrogenation of aliphatic ketones (XVII) is generally similar to the one we have established for incomplete hydrogenation of acetone (XIV), and the kinetic equation (1) is applicable in describing the kinetics of octanone-2 hydrogenation [42].

4. Kinetics of Hydrogenation of Aliphatic Nitriles

Heterogeneous catalytic hydrogenation of nitriles is one of the most economical and widespread methods of synthesis of aliphatic amines, which are valuable products used to make synthetic fibers, detergents, plastics, corrosion inhibitors, medicinal drugs, etc. Acetonitrile has been chosen as the main object of investigation because vapor-phase heterocatalytic hydrogenation of this compound in reactions (III)–(V), albeit a model process, at the same time reflects the typical features of this practically important but poorly studied class of reactions of hydrogenation of CN-containing compounds.

Catalytic hydrogenation of acetonitrile is a complicated process including redox and acid–base conversions, the rate and direction of which depend on the catalyst composition and structure. The detailed mechanism of concurrent syntheses (III)–(V) has heretofore been obscure [43]. Such multipath reactions can most naturally be approached by resolving the complex process into simpler ones. In the case considered such a relatively simple reaction is reaction (III), i.e. synthesis of a primary amine not complicated by side processes of condensation or deamination {43,44] leading to secondary and tertiary amines (reactions IV,V).

A study of the kinetics of reaction (III) on the catalysts (Ni, Co, Cu) displaying one-hundred-percent selectivity for primary amine has shown [45] kinetic patterns similar to those of acetone hydrogenation (I). Figure 6 exemplifies (for a Ni/MgO catalyst) the dependences of the rate of reactions (III) on acetonitrile (P_{CN}) and hydrogen concentrations.

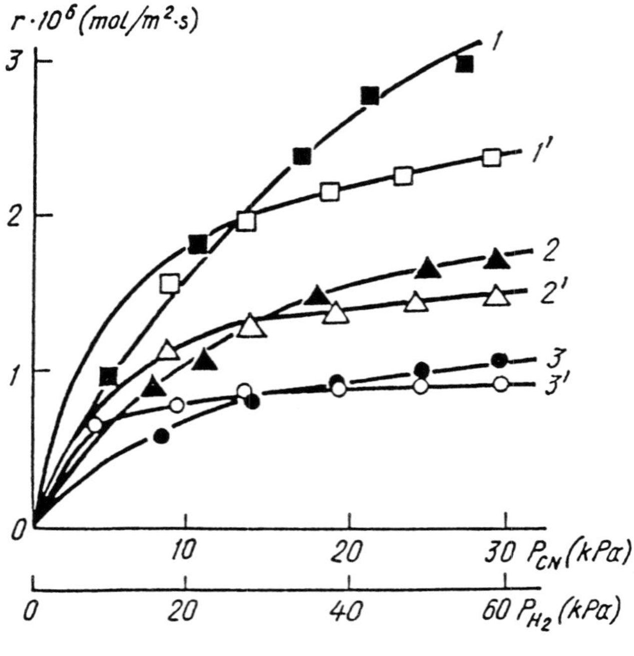

Figure 6. Dependences of the rate of reaction (III) on (1, 2, 3,) P_{CN} and (1', 2', 3') P_{H_2} in the presence of Ni/MgO at (1, 1') 393°K, (2, 2') 378°K, and (3, 3') 369°K.

The forms of acetonitrile adsorption are determined by the structural peculiarities of the CN group. The triple bond $-C\equiv N$ in the acetonitrile molecule is formed by one σ and two π bonds, and is strongly polarized: $-C^{\sigma+}\equiv N^{\sigma-}$[45] because of the greater electronegativity of N (3.04) as compared with C (2.55). Although the nitrogen atom has a free electron pair, it is only weakly basic, because its electron–donor ability is attenuated in an unsaturated compound (CH_3CN). The length of the $C\equiv N$ bond in acetonitrile is 0.116 nm, and the diagonal hybridization of the carbon atom of this group results in the single and the triple bonds being in line [45].

In adsorption on transition metals (M), CH_3CN may act either as a σ–donor forming a linear structure $CH_3-C\equiv N{:}\rightarrow M$, or as a π–acceptor with the electrons of the metal occupying the antibonding π levels of the CN group [46].

On the basis of adsorption [46] and kinetic [45] data the following mechanism has been proposed [45] for reaction (II) on nickel and cobalt catalysts:

$$\begin{aligned}
&(1)\ H_2 + 2Z_H = 2\,Z_H H\\
&(2)\ CH_3CN + Z_N = Z_N CH_3CN\\
&(3)\ Z_N CH_3CN + Z_H H \rightarrow Z_N CH_3CNH + Z_H\\
&(4)\ Z_N CH_3CNH + Z_H H \rightarrow Z_N CH_3CNHN + Z_H\\
&(5)\ Z_N CH_3CHNH + Z_H H \rightarrow Z_N CH_3CH_2NH + Z_H\\
&(6)\ Z_N CH_3CH_2NH + Z_H H \rightarrow CH_3CH_2NH_2 + Z_H + Z_N
\end{aligned} \qquad \text{(XVIII)}$$

It is supposed that dissociative adsorption of hydrogen takes place on single M-sites Z_H, whereas acetonitrile and its partial hydrogenation products bind to multiatom sites Z_N. The first two steps reach an equilibrium, and the limiting step is the interaction of the adsorbed compound with $Z_H H$. Further, several variants are possible. In principle, process (III) may be limited by the rate of one of the four surface reactions (3) to (6).

In the first two steps of hydrogen addition, one or two acetonitrile–metal bonds are split. After addition of two hydrogens, the intermediate acetonitrile hydrogenation product is aldimine $CH_3-CH=NH$, in which carbon is trigonally hybridized and forms one σ and one π bond with the similarly hybridized nitrogen [45]. Polarization of the aldimine double bond of the type $C^{\sigma+}-N^{\sigma-}$ (length 0.124 nm) results in that the carbon–nitrogen bond here to a large extent has the character of a saturated bond. In this case the most probable form of association between the aldimine and the transition metal is the σ–complex

$CH_3CH = NH{:}\rightarrow M$. Therefore one can believe that the adsorbed aldimine and the ensuing CH_3–CH_2–HN: $\rightarrow$ M intermediate are more reactive towards Z_HH than the adsorbed nitrile and the Z_NCH_3CNH particle. Hence it follows that process (I) may be limited by the rate of addition of the first or the second hydrogen atoms; in this case steps (5) and (6) are assumed to be "instant" as defined by Temkin. If step (3) is the limiting one, the kinetic equation is

$$r = k_3 \frac{b_{CN} P_{CN}}{1 + b_{CN} P_{CN}} \cdot \frac{\sqrt{b_{H_2} P_{H_2}}}{1 + \sqrt{b_{H_2} P_{H_2}}} \tag{21}$$

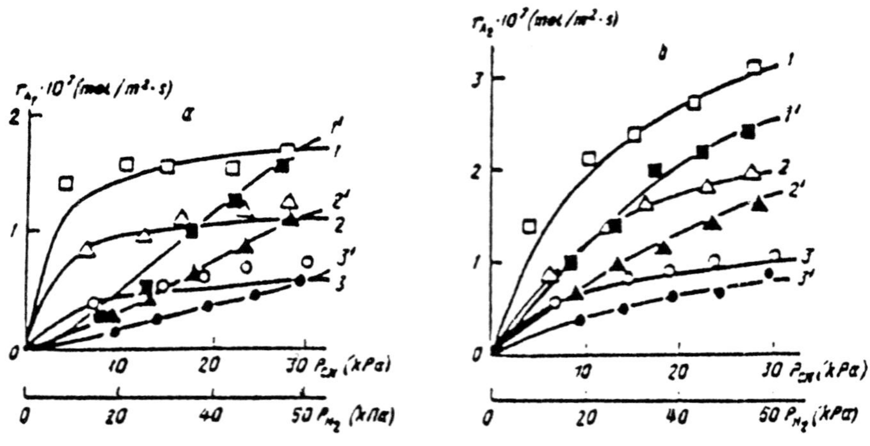

Figure 7. Dependences of the rate of acetonitrile hydrogenation in the presence of Ir black to (a) primary, (b) secondary, and (c) tertiary amines, and of the overall acetonitrile conversion rate (r) in the presence of Ni/MgO on (1, 2, 3) P_{CN} and (1', 2', 3') P_{H_2} at (1, 1') 393°K, (2, 2') 378°K, and (3, 3') 369°K.

where b_{CN} is the nitrile adsorption coefficient. If the rate of the process is determined by the addition of the second hydrogen while step (3) reaches an equilibrium, the kinetic equation becomes

$$r = k_4 \frac{K_3 b_{CN} P_{CN} \sqrt{b_{H_2} P_{H_2}}}{1 + b_{CN} P_{CN} \cdot \left(1 + K_3 \sqrt{b_{H_2} P_{H_2}}\right)} \cdot \frac{\sqrt{b_{H_2} P_{H_2}}}{1 + \sqrt{b_{H_2} P_{H_2}}} \qquad (22)$$

Equation (21) but not (22) is in good correspondence with the experimental data; in Figure 6, the curves have been generated with equation (21), and points are the experimental data.

The kinetics of acetonitrile hydrogenation on a copper catalyst is nicely described by an equation corresponding to the mechanism that implies molecular hydrogen adsorption [45]:

$$r = k_3 \frac{b_{CN} P_{CN}}{1 + b_{CN} P_{CN}} \cdot \frac{b_{H_2} P_{H_2}}{1 + b_{H_2} P_{H_2}} \qquad (23)$$

We have studied [47] the mechanism and kinetics of syntheses (III)–(V) taking place concurrently in the presence of iridium catalysts (Ir black and Ir/Al_2O_3) in the temperature range 371–403K where the influence of the reverse reactions of amine dehydrogenation can be neglected. The rates of acetonitrile conversion to individual products $r_{A1,}$ r_{A2} r_{A3} are expressed as moles $A_i/\sec \cdot m^2$ of metal surface; the overall process rate is $r = r_{A1} + r_{A2} + r_{A3}$, selectivity $S_{Ai} = r_{Ai}/r$. In all cases the reaction rate increases with the concentrations of both reagents (Figure 7).

In formulating the mechanism of the multipath reaction of hydrogenation of acetonitrile and other nitriles, the most complicated question is the one of the chemical nature of the steps giving rise to secondary and tertiary amines. Amine synthesis is believed [43,44] to include, as an intermediate stage, partial hydrogenation of nitrile to aldimine which in some cases [44,48] can be preparatively isolated:

$$CH_3 - C \equiv N + H_2 \rightarrow CH_3 - \underset{(1)}{CH} = NH \xrightarrow{+H_2} CH_3 - \underset{(2)}{CH_2} - NH_2$$

Condensation of aldimine (**1**) and ethylamine (**2**) gives a Schiff base (**3**) which is hydrogenated to diethylamine (**4**):

$$CH_3 - CH = NH = CH_3 - CH_2 - NH_2 \xrightarrow[-NH_3]{} CH_3 - CH = \underset{(3)}{N} - CH_2 - CH \xrightarrow{+H_2}$$

$$\rightarrow \left(CH_3 \underset{(4)}{-} CH_2\right)_2 NH$$

Reaction of aldimine (**1**) with diethylamine (**4**) yields the tertiary amine (**5**):

$$CH_3-CH=NH+(CH_3-CH_2)_2NH \rightarrow CH_3-C(NH_2)H-N(CH_2-CH_3)_2 \rightarrow$$

$$\xrightarrow[-NH_3]{} CH_2=CH-N\left(CH_2-CH_3\right)_2 \xrightarrow{+H_2} N(CH_2 \underset{(5)}{-} CH_3)_3$$

According to such chemistry, aldimine is the common intermediate giving rise to all three amines. Preferential formation of one of the amines would depend on the rates of competing reactions–aldimine hydrogenation and its hydro-condensation with amine.

An alternative point of view [43,48] is that secondary and tertiary amines arise by deamination:

$$2CH_3CH_2NH_2=(CH_3-CH_2)_2NH+NH_3$$

$$CH_3CH_2NH_2+(CH_3-CH_2)_2NH=(CH_3-CH_2)_3N+NH_3$$

or

$$3CH_3=CH_2NH_2=\left(CH_3-CH_2\right)_3N+2NH_3$$

This assertion can be easily checked experimentally. Thus when ethylamine or a mixture of ethyl- and diethylamine are passed over an Ir catalyst at catalytic temperature, no secondary or tertiary amines are formed.

Therefore we have undertaken a computer analysis of about 40 versions of the "condensation" mechanism of amine synthesis. Such analysis demonstrates that the whole body of kinetic and adsorption data obtained can be explained on the basis of the following mechanism of acetonitrile hydrogenation [47]:

	N (III)	N(IV)	N(V)
(1) $H_2 + 2Z_H = 2Z_HH$	2	4	6
(2) $CH_3CN + Z_N = Z_NCH_3CN$	1	2	3
(3) $Z_NCH_3CN + Z_HH = Z_NCH_3CNH + Z_H$	1	1	1
(4) $Z_NCH_3CNH + Z_HH \rightarrow Z_NCH3CHNH + Z_H$	1	1	1
(5) $Z_NCH_3CHNH + Z_HH \rightarrow Z_NCH_3CH_2NH+Z_H$	1	0	0
(6) $Z_NCH_3CH_2NH + Z_HH \rightarrow Z_NA_1 + Z_H$	1	0	0
(7) $Z_NA_1 \rightarrow A_1 + Z_N$	1	0	0
(8) $Z_NCH_3CN + Z_NCH_3CHNH \rightarrow Z_NX_O + Z_N$	0	1	1
(9) $Z_NX_O + Z_HH\ Z_NX_o' \xrightarrow{Z_HH} Z_NA_2 + NH_3 + Z_H$	0	1	0

Continued

(10) $Z_NA_2 \rightarrow A_2 + Z_N$	0	1	0
(11) $Z_NCH_3CN + Z_NX_O \rightarrow Z_NX_1 + Z_N + Z_H$	0	0	1
(12) $Z_NX_1 + Z_HH \rightarrow Z_NX_1' \xrightarrow{Z_HH} Z_NA_3 + NH_3 + Z_H$	0	0	1
(13) $Z_NA_3 \rightarrow A_3 + Z_N$	0	0	1

(XIX)

In general form, mechanism (XIX) includes 30 steps involving 27 independent intermediates, so in this case the Horiuti rule is fulfilled [49] and the reaction proceeds via three kinetically independent routes N(III), N(IV), and N(V) which are summarized by the overall chemical equations (III) to (V).

The following kinetic equations rise from mechanism (XIX) for the rates of formation of individual amines:

$$r_{A1} = r_5 = k_5\Theta_H\Theta_{CHNH}$$
$$r_{A2} = 2r_9 = 2k_9\Theta_H\Theta_{Xo}$$
$$r_{A3} = 3r_{11} = 3k_{11}\Theta_{CN}\Theta_{Xo}$$

where Θ_ι are the occupancies of the Z_H and Z_N sites by the corresponding intermediates, hydrogen, and acetonitrile. The final forms of these equations are:

$$r_{A1} = \frac{k_4K_3b_{CN}P_{CN}\left(b_{H_2}P_{H_2}\right)^{3/2}\Theta^o_{CN}\left(\Theta^o_H\right)^2}{k_5\Theta^o_H\sqrt{b_{H_2}P_{H_2}} + k_8\Theta^o_H b_{CN}P_{CN}} \tag{24}$$

$$r_{A2} = \frac{k_4k_8K_3\left(b_{CN}P_{CN}\right)^2\left(b_{H_2}P_{H_2}\right)^{3/2}\left(\Theta^o_H\Theta^o_{CN}\right)}{\left(k_5\Theta^o_H\sqrt{b_{H_2}P_{H_2}} + k_8\Theta^o_H b_{CN}P_{CN}\right)\cdot\left(k_9\Theta^o_H\sqrt{b_{H_2}P_{H_2}} + k_{11}\Theta^o_H b_{CN}P_{CN}\right)} \tag{25}$$

$$r_{A3} = \frac{k_4k_8K_3\left(b_{CN}P_{CN}\right)^3\left(b_{H_2}P_{H_2}\right)\Theta^o_H\left(\Theta^o_{CN}\right)^3}{\left(k_5\Theta^o_H\sqrt{b_{H_2}P_{H_2}} + k_8\Theta^o_H b_{CN}P_{CN}\right)\cdot\left(k_9\Theta^o_H\sqrt{b_{H_2}P_{H_2}} + k_{11}\Theta^o_H b_{CN}P_{CN}\right)} \tag{26}$$

The portion of free Z_H sites is

$$\Theta_{\mathrm{H}}^{\mathrm{o}} = 1\left(\sqrt{b_{\mathrm{H}_2} P_{\mathrm{H}_2}}\right)^{-1}$$

whereas the portion of free Z_N sites is obtained by solving the cubic equation

$$a\left(\Theta_{\mathrm{CN}}^{\mathrm{o}}\right)^3 + b\left(\Theta_{\mathrm{CN}}^{\mathrm{o}}\right)^2 + c\Theta_{\mathrm{CN}}^{\mathrm{o}} + d = 0$$

where

$$a = k_8 k_{11} b_{\mathrm{CN}} P_{\mathrm{CN}}\left[1 + b_{\mathrm{CN}} P_{\mathrm{CN}}\left(1 + K_3\sqrt{b_{\mathrm{H}_2} P_{\mathrm{H}_2}}\right)\right]$$

$$b = b_{\mathrm{CN}} P_{\mathrm{CN}}\left\{\sqrt{b_{\mathrm{H}_2} P_{\mathrm{H}_2}}\,\Theta_{\mathrm{H}}^{\mathrm{o}}\left(k_5 k_{11} + k_8 k_9\right)\left[1 + b_{\mathrm{CN}} P_{\mathrm{CN}}\left(1 + K_3\sqrt{b_{\mathrm{H}_2} P_{\mathrm{H}_2}}\right)\right] + \right.$$
$$\left. + b_{\mathrm{CN}} P_{\mathrm{CN}}\left[k_4 K_3 b_{\mathrm{H}_2}\Theta_{\mathrm{H}}^{\mathrm{o}}\left(k_8 + k_{11}\right) - k_8 k_{11}\right]\right\}$$

$$c = \frac{k_9 b_{\mathrm{H}_2} P_{\mathrm{H}_2}\left(\Theta_{\mathrm{H}}^{\mathrm{o}}\right)^2\left\{k_4 K_3 b_{\mathrm{CN}} P_{\mathrm{CN}}\sqrt{\left(b_{\mathrm{H}_2} P_{\mathrm{H}_2}\right)} + k_5\left[1 + b_{\mathrm{CN}} P_{\mathrm{CN}}\left(1 + K_3\sqrt{b_{\mathrm{H}_2} P_{\mathrm{H}_2}}\right)\right]\right\} -}{-b_{\mathrm{CN}} P_{\mathrm{CN}}\Theta_{\mathrm{H}}^{\mathrm{o}}\sqrt{b_{\mathrm{H}_2} P_{\mathrm{H}_2}}\left(k_5 k_{11} + k_8 k_9\right)}$$

$$d = -k_9 k_5 b_{\mathrm{H}_2} P_{\mathrm{H}_2}\left(\Theta_{\mathrm{H}}^{\mathrm{o}}\right)^2$$

Equations (24) to (26) are in quantitative correspondence with the experimental kinetic dependences (in Figure 7) the curves have been obtained from the equations, and points are the experimental data) and describe the whole body of the experimental results. The mechanism proposed for acetonitrile hydrogenation to primary, secondary, and tertiary amines is confirmed not only by the agreement between the observed kinetic dependences and the kinetic equations (24) to (26), but also by the fact that the constants in these equations calculated from the experimental data have reasonable physical meaning conforming to the nature of relevant steps in mechanism (XIX) [47].

Vapor-phase hydrogenation of other aliphatic mononitriles is far less studied than that of acetonitrile. The scanty data available point to a certain similarity in the behavior of different nitriles during their reduction. Thus the activity of metal catalysts in the reaction of dinitrile hydrogenation decreases in the order Pt, Pd > Ni > Co > Cu, Fe [43,44] which is generally similar to the series that we have obtained for acetonitrile [50]. It can therefore be supposed that the patterns established for acetonitrile hydrogenation are typical enough of this class of reactions.

This is evidenced by the fact that using the notions presented above it is possible to interpret the order of relative reactivity of substituted nitriles R-CN (where R is a substituent with one to four carbon atoms) which we have found for a nickel catalyst [51]: $-CH_3 > -C_2H_5 > -C_3H_7 > -C_4H_9$. Such a series can be explained on the basis of mechanism (XIX) from which it follows that at rather high extents of occupation of the surface by the nitrile (Θ) exceeding the optimal value Θ_{opt} the catalytic activity r must decrease with increasing R-CN adsorption heat (q_R). At low occupancy ($\theta < \theta_{opt}$) the activity would decrease with decreasing q_R and so in the general case the dependence of r on q_R passes through a maximum.

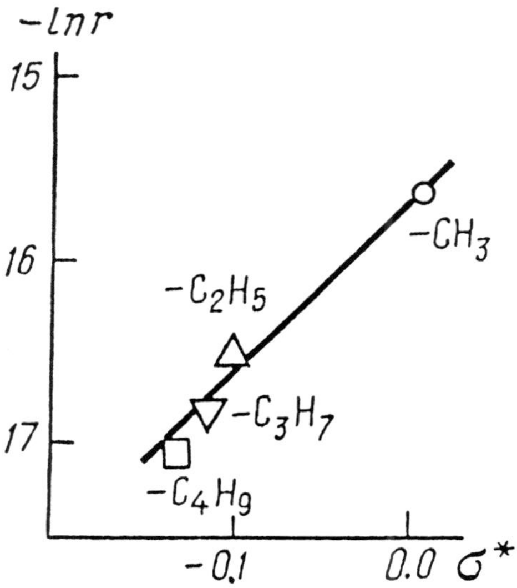

Figure 8. Dependences of the rate of hydrogenation of nitriles R-CN on Taft σ^* constants. R: (1) $-CH_3$, (2) $-C_2H_5$, (3) $-C_3H_7$, and (4) $-C_4H_9$.

For hydrogenation of nitriles of different structure on one and the same catalyst, the adsorption heat and therefore the reaction rate depend on the nitrile structure. With increasing number of carbons in the alkyl substituent *R* the electron density in the CN group is enhanced owing to the induction effect. The same order is found for the decrease in the Taft σ^* constants for alkyl substituents [52], i.e. for the increase in the ability of the alkyl groups to repel electrons towards the CN group (Figure 8).

The induction effect fades along a chain of simple bonds, but there is a steric factor acting in the same direction: the larger is R the smaller is the number of the Z_N and Z_H sites capable of adsorbing the nitrile and hydrogen. Thus valeronitrile is hydrogenated more slowly than butyronitrile despite the same electron density on the CN group [51].

5. Kinetics of Hydrogenation of Carbon Monoxide and Dioxide

5.1 Hydrogenation of CO

In a number of studies on the kinetics of carbon monoxide hydrogenation [52–58] the reaction rate dependence on partial tensions of the components has been expressed by an exponential equation:

$$r = -k \cdot P_{CO}^{m} P_{H_2}^{n} P_{H_2O}^{l} \qquad (27)$$

where m, n, and l are the reaction orders and P_i are the current partial tensions of the reagents. Equation (27) is of course an approximation of the more complex kinetic dependences and does not reflect the actual mechanism of the process. Equations of this type are only valid in a limited range of changes in the process parameters, and their constants often have no physical meaning. Therefore here such works will not be considered any further.

According to Herwijnen et al. [59], the dependence of the hydrogenation rate (Ni/Al_2O_3, 443–483°K) on CO concentration passes through a maximum. The kinetic data obtained in a flow system are described by an equation

$$r = kP_{CO}/(1 + k_1 P_{CO})^2$$

(k_1 is the equilibrium constant of CO adsorption) that is in accord with the scheme proposed by Vlasenko and Yuzefovich [57]:

(1) $H_2 + Z = ZH_2$

(2) $CO + Z = ZCO$

(3) $ZCO + ZH_2 = ZHCOH + Z$

$$(4)\ ZHCOH + ZH_2 = ZCH_2 + H_2O + Z \qquad (XX)$$

$$(5) ZCH_2 + ZH_2 = CH_4 + 2Z$$

when step (4) is the limiting one. However, Herwijnen et al. [59] do not take into account the influence of hydrogen concentration on the reaction rate. A more complete equation corresponding to mechanism (XX) and describing the rate of reaction (VII) on some metals of the VIIth group at 503–528°K is given by Vannice [60]:

$$r = \frac{k_1 P_{H_2} P_{CO}}{\left(1 + k_2 P_{H_2} + k_3 P_{CO} + k_4 P_{H_2} P_{CO}\right)^2}$$

The denominator in this equation reflects the presence of different intermediates of the CHOH type on the catalyst surface.

The kinetic equation given by Huang and Richardson [61] for CO hydrogenation on nickel catalysts

$$r = k P_{CO} \sqrt{P_{H_2}} \Big/ \left(1 + k_1 P_{CO}\right)^2$$

under certain assumptions (the third step is limiting, the surface is uniform and occupied by nondissociated CO molecules) stems from the following scheme:

$$(1)\ H_2 + 2Z = 2ZH$$

$$(2)\ CO + Z = ZCO$$

$$(3)\ ZCO + ZH = ZCOH + Z$$

$$(4)\ ZCOH + ZH = ZHCOH + Z$$

$$(5)\ ZHCOH + ZH = ZH_2COH + Z \qquad (XXI)$$

$$(6) ZH_2COH + ZH = ZCH_2 + H_2O + Z$$

$$(7)\ ZCH_2 + ZH = ZCH_3 + Z$$

$$(8)\ ZCH_3 + ZH = CH_4 + 2Z$$

Studies of reaction (VII) on Ni/SiO_2 at 473–603K by nonsteady-state methods, based on the scheme

$$\begin{aligned}
&(1)\ CO + Z = ZCO \\
&(2)\ H_2 + 2Z = 2ZH \\
&(3)\ ZCO + 2ZH = ZC + H_2O + 2Z \\
&(4)\ ZC + 4ZH = CH_4 + 5Z
\end{aligned} \qquad \text{(XXII)}$$

and on the assumption that step (3) is slow, yielded [62] the following kinetic equation:

$$r = kP_{CO}P_{H_2} \Big/ \left(1 + k_1P_{CO} + k_2P_{H_2}\right)^2 \qquad (28)$$

Scheme (XXII) and equation (28), in contrast to mechanisms (XX) and (XXI), imply dissociation of the CO molecule on the catalyst surface and formation of a surface carbide.

The kinetic data obtained with an industrial Ni/Al_2O_3 catalyst [63] correspond, in the authors' opinion, to the following scheme:

$$\begin{aligned}
&(1)\ CO + 2Z = ZC + ZO \\
&(2)\ H_2 + 2Z = 2ZH \\
&(3)\ ZC + 2ZH = ZCH_2 + 2Z \\
&(4)\ ZCH_2 + 2ZH = CH_4 + 3Z \\
&(5)\ ZO + 2ZH = H_2O + 3Z
\end{aligned} \qquad \text{(XXIII)}$$

Mechanism (XXIII) implies dissociative adsorption of CO with ensuing hydrogenation of the surface carbon and ZO. Assuming that step (3) is limiting, the authors arrive at the following kinetic equation:

$$r = \frac{k_1\sqrt{P_{CO}}\,P_{H_2}}{\left(1 + k_2\sqrt{P_{CO}} + k_3\sqrt{P_{H_2}}\right)^3}$$

For a mechanism similar to (XXIII) in implying dissociation CO adsorption but more detailed in respect of hydrogen addition to ZC, Meerten et al. [64] have obtained the followinf model:

$$r = \frac{k_1 \sqrt{P_{CO}} P_{H_2}}{\left(1 + k_2 \sqrt{P_{H_2}} + k_3 \sqrt{P_{CO}} + k_4 \sqrt{P_{H_2} P_{CO}}\right)^2}$$

Mention should also be made of the works of Ibraeva et al. [65, 66] on the kinetics of complete reduction of CO on nickel catalysts, where it has been shown that the experimental results are best described by the equation

$$r = \frac{k_1 \sqrt{P_{CO}} P_{H_2}}{\sqrt{P_{H_2}} + k_2 P_{CO} + k_3 \sqrt{P_{H_2} P_{CO}}}$$

This equation is interpreted [65,66] on the basis of the following mechanism:

$$\begin{aligned}
&(1)\ H_2 + 2Z = 2ZH \\
&(2)\ CO + Z = ZCO \\
&(3)\ ZCO + ZH = ZCOH + Z \\
&(4)\ ZCOH + ZH = ZHCOH + Z \\
&(5)\ ZHCOH + ZH = ZCH + ZOH \\
&(6)\ ZCH + ZH = ZCH_2 + Z \\
&(7)\ ZCH_2 + ZH = ZCH_3 + Z \\
&(8)\ ZCH_3 + ZH = CH_4 + 2Z \\
&(9) ZOH + ZH = H_2O + 2Z
\end{aligned} \qquad \text{(XXIV)}$$

It is assumed that step (6) is limiting and that the catalyst surface is nonuniform in respect of reagent adsorption energy.

Numerous studies have been devoted to CO hydrogenation on metals leading mainly to methane and its homologs. Yet, the proposed mechanisms only pertain to methane formation, and ignore that the rate of reaction (VII) can be affected by concurrent formation of other hydrocarbons sharing common steps with CH_4.

We have studied the kinetics of reactions (VII) and (VIII) on metal–oxide catalysts differing in the chemical nature of the metal and the composition of the carrier [67–73]. The data on the kinetics of CO hydrogenation in the presence of Ni/TiO_2 are presented in Figure 9.

Analogous typical dependences have been obtained with other catalysts tested: the overall CO conversion rate (r) increases linearly with P_{H_2}; the dependence of r on P_{CO} passes through a maximum.

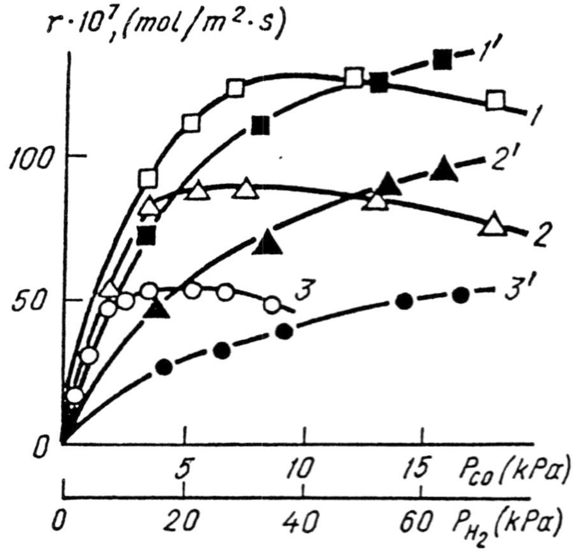

Figure 9. Dependences of the rate of CO hydrogenation in the presence of Ni/TiO_2 on (1, 2, 3) P_{CO} and (1', 2', 3') P_{H_2} at (1, 1') 561°K, (2, 2') 545°K, and (3, 3') 526°K.

The adsorption –desorption assays for a nickel catalyst that we have carried out [74] using thermogravimetric, mass spectrometric, and IR-spectroscopic techniques allow a conclusion that on the hydrogen-saturated surface of the metal CO is adsorbed in molecular form.

In the aggregate, the adsorption and kinetic data suggest that CO and atomic hydrogen interact in the adsorbed state, and the extremal character of the $r = f(P_{CO})$ curves is indicative of reagent adsorption on the same sites of the catalyst surface. The mechanism of the process can be depicted with the following scheme [71,72]:

(1) $H_2 + 2Z = 2ZH$

(2) $CO + Z = ZCO$

(3) $ZCO + ZH = ZCOH + Z$ (XXV)

(4) $ZCOH + ZH \rightarrow ZX_O + Z$

(5) $ZX_O + ZH \rightarrow ZX_o' \xrightarrow{ZH} CH_4 + H_2O + Z$

(6) $ZX_O + ZY \rightarrow ZX_1 \xrightarrow{ZH} C_2H_6 + ... + Z$

In mechanism (XXV) steps (1) to (4) determine the activity, and steps (5) and (6) the selectivity of the process. Equilibrium is attained in steps (1) to (3), and the limiting step is the surface reaction between ZCOH and ZH. The adsorbed particle ZX_O, the composition of which can be expressed as ZCHOH, is the common intermediate for all products. Its interaction with ZH yield methane (chain termination) whereas its reaction with the carbon-containing particle ZY (ZCO, ZCHOH, or ZCH_2) is the onset of a hydrocondensation process yielding the compound ZX_1 leading to higher hydrocarbons (chain growth).

Mechanism (XXV) gives rise to the following kinetic equation of overall CO conversion [67, 68, 75]:

$$r = k_4 \frac{k_3 b_{CO} P_{CO} b_{H_2} P_{H_2}}{\left[1 + b_{CO} P_{CO} + \sqrt{b_{H_2} P_{H_2}} \left(1 + K_3 b_{CO} P_{CO}\right)\right]^2} \tag{29}$$

A version of scheme (XXV) is the mechanism where the third step is limiting:

(1) $H_2 + 2Z = 2ZH$

(2) $CO + Z = ZCO$

(3) $ZCO + ZH \rightarrow ZCOH + Z$ (XXVI)

In this case the kinetics of the overall process is described by the equation

$$r = k_3 \frac{b_{CO} P_{CO} \sqrt{b_{H_2} P_{H_2}}}{\left(1 + b_{CO} P_{CO} + \sqrt{b_{H_2} P_{H_2}}\right)^2}$$

If the carbon-containing intermediates bind with two Z sites, mechanism (XXV) is modified to

(1) $H_2 + 2Z = 2ZH$

(2) $CO + Z = ZCO$

$$(3)\ ZCO + ZH = Z_2COH \qquad (XXVII)$$

$$(4)\ Z_2COH + ZH \rightarrow Z_2X_O + Z$$

Provided that the concentration of the Z_2CHO particles is negligible, the CO conversion rate is described by the equation

$$r = k_4 \frac{K_3 b_{CO} P_{CO} b_{H_2} P_{H_2}}{\left(1 + b_{CO} P_{CO} + \sqrt{b_{H_2} P_{H_2}}\right)^3}$$

If we assume that two adsorbed hydrogen atoms react simultaneously in the limiting step [i.e., steps (3) and (40) of mechanism (XXV) are united in one: (3') $ZCO + 2ZH \rightarrow ZX_O + 2Z$], the kinetic equation acquires the form

$$r = k_3 \frac{b_{CO} P_{CO} b_{H_2} P_{H_2}}{\left(1 + b_{CO} P_{CO} + \sqrt{b_{H_2} P_{H_2}}\right)^3}$$

Equation (29) can be modified somewhat. From the condition of the maximum on the $r = f\,(P_{CO})$ curve we obtain

$$b_{CO} P_{CO}^{max} \frac{1 + \sqrt{b_{H_2} P_{H_2}}}{1 + K_3 \sqrt{b_{H_2} P_{H_2}}} \qquad (30)$$

so that

$$r^{max} = \frac{k_4 K_3 b_{H_2} P_{H_2}}{4\left(1 + \sqrt{b_{H_2} P_{H_2}}\right) \cdot \left(1 + K_3 \sqrt{b_{H_2} P_{H_2}}\right)} \qquad (31)$$

and by substituting (30), (31) into equation (29) we obtain

$$r = 4r^{max} \cdot P_{CO}^{max} \cdot P_{CO} \Big/ \left(P_{CO}^{max} + P_{CO}\right)^2 \qquad (32)$$

Equation (32) provides a quantitative description of the experimental data for all catalysts that we have studied and is also in accord with other data [76,77] at the values of the constants meeting conditions (30) and (31).

The mechanism (XXV) that we have established for CO hydrogenation and the kinetic model (29) have some important consequences.

Figure 10 presents the temperature dependence of the rate of CO overall conversion on some metal catalysts. As can be seen, *r* increases with temperature, reaching a plateau. Outwardly this resembles the transition of the reaction from the kinetic domain to that of internal and then external diffusion; yet, such a transition was ruled out in our experiments. The cause of such a pattern of the $r = f$ (T) dependence lies in the substantial changes in the catalyst occupancy by the reagents that take place with increasing temperature.

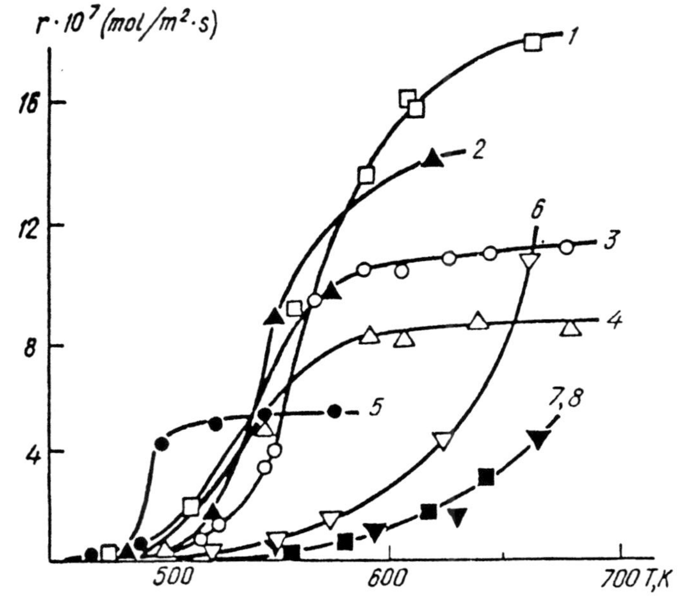

Figure 10. Temperature dependences of the CO hydrogenation rates with metal/Al_2O_3 catalysts: (1) Ru, (2) Rh, (3) Co, (4) Ni, (5) Ni/Cr_2O_3, (6) Pt., (7) Pd, (8) Ir.

In accordance with equation (29) we obtain the following expression for the apparent activation energy determined from the $r = f$ (T) dependence at constant P_i [69, 80]:

$$E = -R\left(\frac{d\ln r}{d(1/T)}\right)_{P_i}$$

$$= E_4 - Q_3 + (2\Theta_{CO} - 1)q_{CO} - 1 + (\Theta_H - 1)q_{H_2} \quad (33)$$

Considering the observed first order of the reaction in H_2 ($\Theta_H << 1$), equation (33) can be simplified to

$$E = E_4 - Q_3 + (2\Theta_{CO} - 1)q_{CO} - q_{H_2} \quad (34)$$

At high temperatures $\Theta_{CO} \to O$ and E tends to the minimal value

$$E \to E_{min} = (E_4 - Q_3) - q_{CO} - q_{H_2} \quad (35)$$

At low temperatures $\Theta_{CO} \to 1$ and E tends to the maximal value

$$E \to E_{max} = (E_4 - Q_3) + q_{CO} - q_{H_2} \quad (36)$$

At medium occupancies $(\Theta_{CO} \approx 0.5)$ we have

$$E = (E_4 - Q_3) - q_{H_2} \quad (37)$$

Expressions (35) to (37) can be used to estimate the values ($E_4 - Q_3$) and Θ_{CO}; from these relations it follows that

$$q_{CO} = 0.5(E_{max} - E_{min}) \quad (38)$$

Table 1 shows that q_{CO} values found from equation (38) agree with those obtained from the temperature dependence of b_{CO}.

Table 1. Adsorption of CO on metal catalysts

	q_{CO}, kJ/mole	
Catalyst	From equation (38)	From the temperature dependence of b_{CO}
CO/Al_2O_3	72	70
Ni/Al_2O_3	91	58
Ru/Al_2O_3	54	53
Rh/Al_2O_3	82	78

If step (6) of mechanism (XXIII) is envisaged as

$$(6)\ \mathrm{ZCHOH} + \mathrm{ZCHOH} \rightarrow \mathrm{ZX_1} \xrightarrow{\mathrm{ZH}} \mathrm{C_2H_6} + \ldots + \mathrm{Z}$$

i.e., the synthesis of methane homologs begins with condensation of two ZX_O particles ($ZY = ZX_O$), we obtain the following expression for the selectivity for methane, S_{CH_4} (ignoring the concentrations of intermediates yielding hydrocarbons of more than two carbon atoms, because of the negligible content of C_3, C_4 paraffins in the reaction products) [71,72]:

$$S_{CH_4} = r_5/(r_5 + r_6) = \left(0.5 + \sqrt{0.25 + 2k_4k_6K_3b_{CO}\,P_{CO}/k_5^2}\right)^{-1} \tag{39}$$

or $S_{CH_4} = \left(0.5 + \sqrt{0.25 + cP_{CO}}\right)^{-1}$

where $c = 2k_4k_5^{-2}k_6K_3b_{CO}$

Then the kinetic equation of reaction (VII) can be put down as

$$r = k_4 \frac{K_3b_{CO}P_{CO}b_{H_2}P_{H_2}\left(0.5 + \sqrt{0.25 + cP_{CO}}\right)^{-1}}{\left[1 + b_{CO}P_{CO} + \sqrt{b_{H_2}P_{H_2}}\left(1 + K_3b_{CO}P_{CO}\right)\right]^2}$$

Selectivity for higher hydrocarbons is $S_h = 1 - S_{CH_4}$. Hence

$$S_h = 1 - \left(0.5 + \sqrt{0.25 + cP_{CO}}\right)^{-1} \tag{40}$$

and we obtain the following kinetic equation for the Fischer–Tropsch process (reaction VIII):

$$r = k_4 \frac{K_3b_{CO}P_{CO}b_{H_2}P_{H_2}\left[1 - \left(0.5 + \sqrt{0.25 + cP_{CO}}\right)^{-1}\right]}{\left[1 + b_{CO}P_{CO} + \sqrt{b_{H_2}P_{H_2}}\left(1 + K_3b_{CO}P_{CO}\right)\right]^2} \tag{41}$$

As follows from equations (39) and (40), at P_{CO} = const the selectivity is independent of the hydrogen concentration in the reaction mixture. Besides, equation (39) implies that plotting $A = \left(S_{CH_4}{}^{-1} - 0.5\right)^2$ vs. P_{CO} should yield a straight line. The experimental data presented in Figure 11 confirm these conclusions. Similar dependences have been obtained for other catalysts. Another argument in favor of mechanism (XXIII) with the chain growth step (6) envisaged as $ZX_O + ZX_O \rightarrow ...$ is the fact that the experimental curves in the A vs. P_{CO} plots converge in one point (0.25 on the ordinate axis) coinciding with the value of the ordinate calculated from equation (39).

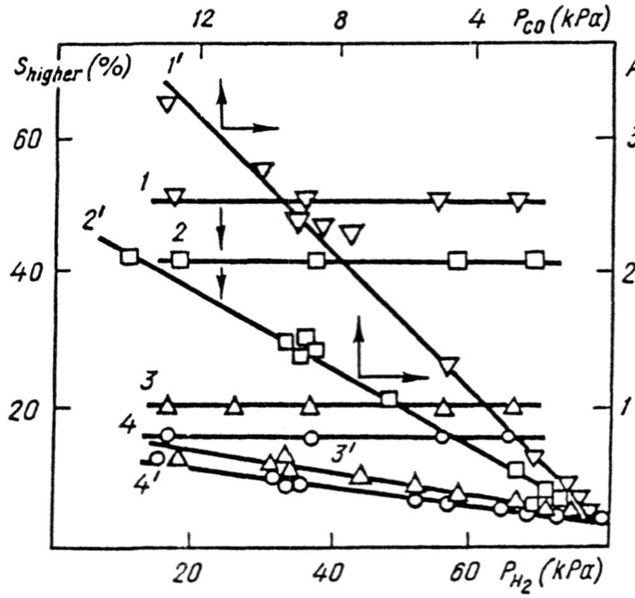

Figure 11. Dependences of the selectivity of CO hydrogenation with nickel catalysts on (1, 2, 3) P_{CO} and (1', 2', 3') P_{H_2}. The carriers are (1, 1') ZrO_2, (2, 2') MgO, (3, 3') TiO_2, (4, 4') SiO_2.

If step (6) is assumed to be

$$(6)\ ZCHOH + ZCO \rightarrow ZX_1 \xrightarrow{ZH} C_2H_6 + \ldots + Z$$

corresponding to intrusion of the CO molecule at the metal–carbon bond, selectivity would be expressed as

$$S_h = \left(1 + k_5 \sqrt{b_{H_2} P_{H_2}} / k_6 b_{CO} P_{CO}\right)^{-1}$$

i.e., would increase with increasing ratio P_{CO}/P_{H_2}, which contradicts the experimental data.

A version of scheme (XXV) is [71,72]

$$\begin{aligned}
&(5)\ ZX_o + ZH \rightarrow ZX_o' + Z \\
&(5')\ ZX_o' + ZH \rightarrow ZX_1 \xrightarrow{ZH} CH_4 + H_2O + Z \\
&(6)\ ZX_1 + ZX_1 \rightarrow ZX_2 \xrightarrow{ZH} C_2H_6 + \ldots + Z
\end{aligned} \tag{XXVIII}$$

where ZX_1 means the intermediate particle ZCH_2. Mechanisms (XXV) and (XXVIII) are kinetically indistinguishable and may supplement each other.

The rate of formation of individual hydrocarbons can be obtained by combining equation (41) with the known distribution [69]:

$$C_n = n(1-\alpha)^2 \cdot \alpha^{n-1} \tag{42}$$

where C_n is the concentration of a hydrocarbon with n carbon atoms; coefficient α is the probability of formation of a higher hydrocarbon, i.e., the ratio of the chain growth rate to the sum of chain growth and termination rates: $\alpha = r_6/(r_5 + r_6)$. Hence the physical meaning of this coefficient: α in distribution (42) is equal to S_h given by equation (40).

5.2 Hydrogenation of CO_2

The mechanism and kinetics of CO_2 hydrogenation are far less studied than those of carbon monoxide. Several exponential equations, with reaction orders in CO_2 and H_2 varying from zero to first, have been proposed to describe the process on nickel [78–82], rhodium [83], and ruthenium [84] catalysts. Such exponential equations do not ensure the reliability of the kinetic information, since they only approximate the more complex actual dependences and are valid in a limited range of parameters.

The kinetics of reaction (IX) in a cycled flow in the presence of Ni/Cr_2O_3 has been examined by Naumov and Krylov [85,86] who have obtained the following kinetic equation:

$$r = k_1 P_{CO_2} P_{H_2} / \left(K_3 + P_{CO_2} + k_2 P_{H_2}\right)$$

where P_{CO_2} is the partial CO_2 tension.

In recent years a number of researchers [87–90] have given preference to a reaction mechanism in which adsorbed carbon dioxide dissociates on the surface to CO or C and O. Thus Maatman and Hiemstra [87] propose the following scheme for reaction (IX) on Ni/Al_2O_3:

$$\begin{aligned}
&(1)\ H_2 + 2Z = 2ZH \\
&(2)\ CO_2 + 2Z = ZCO + ZO \\
&(3)\ ZCO + 2ZH = ZC + H_2O + 2Z \\
&(4)\ ZO + 2ZH = H_2O + 3Z \\
&(5)\ ZC + H_2 = ZCH_2 \\
&(6)\ ZCH_2 + H_2 = CH_4 + Z
\end{aligned} \qquad \text{(XXIX)}$$

However, the authors give no substantiation of this mechanism. The kinetic equation

$$r = k_1 P_{CO_2} \Big/ \left(1 + k_2 P_{CO_2} + k_3\sqrt{P_{H_2}}\right) \qquad (43)$$

obtained by Zakumbaeva et al. [90] for the kinetics of CO_2 hydrogenation on nickel catalysts corresponds to scheme (XXIX) and provides a description of the experimental data in the whole range of the reaction parameters.

In a detailed study of reaction (IX) on Ni/SiO_2 [89], the following mechanism has been put forward:

$$\begin{aligned}
&(1)\ H_2 + 2Z = 2ZH \\
&(2)\ CO_2 + 2Z = ZCO + ZO \\
&(3)\ ZCO = CO + Z \\
&(4)\ ZCO + Z = ZC + ZO \\
&(5)\ ZC + ZH = ZCH + Z \\
&(6)\ ZCH + ZH = ZCH_2 + Z \\
&(7)\ ZCH_2 + ZH = ZCH_3 + Z \\
&(8)\ ZCH_3 + ZH = ZCH_4 + Z \\
&(9)\ ZCH_4 = CH_4 + Z \\
&(10)\ ZO + ZH = ZOH + Z \\
&(11)\ ZOH + ZH = ZH_2O + Z \\
&(12)\ ZH_2O = H_2O + Z
\end{aligned} \qquad \text{(XXX)}$$

On the basis of the data obtained, the authors [89] have proposed a kinetic equation

$$r = k_4 \frac{k_1\sqrt{P_{CO_2}P_{H_2}}}{\left(1+K_2\sqrt{P_{CO_2}}/\sqrt{P_{H_2}}+K_3\sqrt{P_{CO_2}}+K_4P_{CO}\right)^2}$$

which in their opinion corresponds to mechanism (XXX).

Serov et al. [91] have studied the kinetics of CO_2 hydrogenation to CO and methane on Pd–Ru films and considered a 16-step, 6-route mechanism of concurrent reactions (IX) and (XI). Analysis of the kinetic model has allowed the authors to conclude that the process can be described by a mechanism with a smaller number of steps involving only three routes:

	(i)	(ii)	(iii)
(1) $H_2 + 2Z = 2ZH$	1	1	1
(2) $CO_2 + Z = ZCO_2$	1	1	1
(3) $ZCO_2 + 2ZH = Z_3CHOOH$	1	1	1
(4) $Z_3CHOOH = Z_3CO + H_2O$	0	1	1
(5) $ZCO + 2ZH = Z_3CHOH$	0	1	0
(6) $Z_3CHOH + 2ZH = Z_3CH_2 + H_2O + 2Z$	1	1	0
(7) $Z_3CH_2 + 2ZH = CH_4 + 5Z$	1	1	0
(8) $Z_3CHOOH + 2ZH = Z_3CHOH + H_2O + 2Z$	1	0	0
(9) $Z_3CO = CO + 3Z$	0	0	1

(XXXI)

Route (i), i.e. methane synthesis, is actualized at lower temperatures (up to 500K): at higher temperatures CH_4 is formed via route (ii); route (iii) represents reduction of CO_2 to CO. The kinetic equations corresponding to each route are as follows:

$$r_i = \frac{K_i P_{CO_2}\left(P_{H_2}\right)^2}{\left[1+\sqrt{(K_1)^{-1}\cdot P_{H_2}}+K_2P_{CO_2}\right]^2} \tag{44}$$

$$r_{ii} = \frac{K_{ii} P_{CO_2}\left(P_{H_2}\right)^2}{\left[1+\sqrt{(K_1)^{-1}\cdot P_{H_2}}+K_2P_{CO_2}\right]^5 \cdot P_{H_2O}} \tag{45}$$

$$r_{iii} = \frac{K_{iii} P_{CO_2} P_{H_2}}{\left[1 + \sqrt{(K_1)^{-1} \cdot P_{H_2}} + K_2 P_{CO_2}\right]^3 \cdot P_{H_2O}} \tag{46}$$

The authors [91] do not make a detailed analysis of the kinetic models obtained and of the constants entering into equations (44) to (46). Yet, consideration of the kinetic parameters of these equations reveals that the numerical values of some constants do not conform to their physical meaning (thus the adsorption constant K_1 decreases while K_2 increases with temperature). Besides, the work [91] fails to provide a convincing explanation for the temperature-dependent switchover in the hydrogenation mechanism.

Thus, the above presentation demonstrates that the notions on the mechanism and kinetics of CO_2 hydrogenation are equivocal and controversial. On the whole, the kinetics and mechanisms of synthesis (IX)–(XI) are insufficiently studied, as the works available mainly concern the kinetics of reaction (IX).

We have carried out a detailed study [92] of the kinetics of CO_2 hydrogenation on transition metals: Fe, Co, Ni, Cu, Rh, Pd, Ir, Pt—both solid and applied on Al_2O_3—and also Rh applied on TiO_2, SiO_2, MgO, ZrO_2, and Nb_2O_5. Typical rate dependences of CO_2 conversion to corresponding products are shown in Figures 12 and 13.

The key question in elucidating the mechanism of CO_2 hydrogenation is whether hydrocarbon synthesis takes place in series through intermediate formation of CO: $CO_2 \rightarrow CO \rightarrow C_xH_y$, or CH_4 and its homologs are formed according to a "series-parallel" scheme. Most researchers [78, 83–89] note that the overall CO_2 hydrogenation rate is higher than that of CO. A similar result has also been obtained in our studies [92]. This fact is indicative of the different hydrogenation routes for carbon monoxide and dioxide. The same conclusions are made in the works [93,94] pointing out that ZCO and consequently ZC are not intermediates in reaction (IX).

Adsorption of CO_2 on transition metals is practically not studied; on the one hand, direct adsorption assays [57,84,95] show that CO_2 is not adsorbed in Ni or Rh, and on the other, in a mixture with hydrogen CO_2 rapidly and reversibly binds with the transition metal.

The lack of clear notions on the mechanism of adsorption and primary activation of the CO_2 molecule made us undertake an analysis of different kinetic models of reaction (IX) on nickel and rhodium catalysts which give rise to methane only, and a comparison of the calculated data with the experimental dependences.

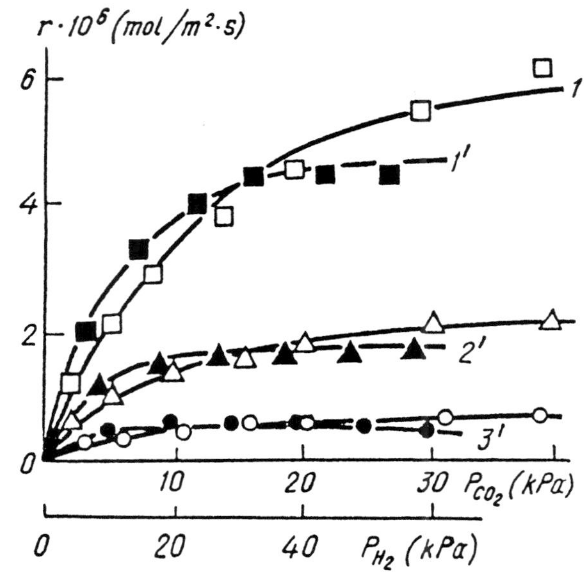

Figure 12. Dependences of the rate of CO_2 hydrogenation in the presence of Ni/Al_2O_3 on (1, 2, 3) P_{CO_2} and (1', 2', 3') P_{H_2} at (1, 1') 534°K, (2, 2') 504°K, and (3, 3') 473°K.

Let us consider a mechanism

$$(1)\ H_2 + 2Z = 2ZH$$
$$(2)\ CO_2 + Z = ZCO_2$$
$$(3)\ ZCO_2 = ZH \rightarrow ZCO_2H + Z$$
$$(4)\ ZCO_2H + ZH \rightarrow ZC(OH)_2 + Z$$
$$(5)\ ZC(OH)_2 + ZH \rightarrow ZCOH + H_2O + Z$$
$$(6)\ ZCOH + ZH \rightarrow ZCHOH \xrightarrow{ZH} \ldots CH_4 + H_2O + Z \qquad \text{(XXXII)}$$

which implies reversible equilibrium adsorption of CO_2 and H_2 on the same sites Z of the metal surface. In this case several variants are possible. If the process is limited by step (3), the corresponding kinetic equation is

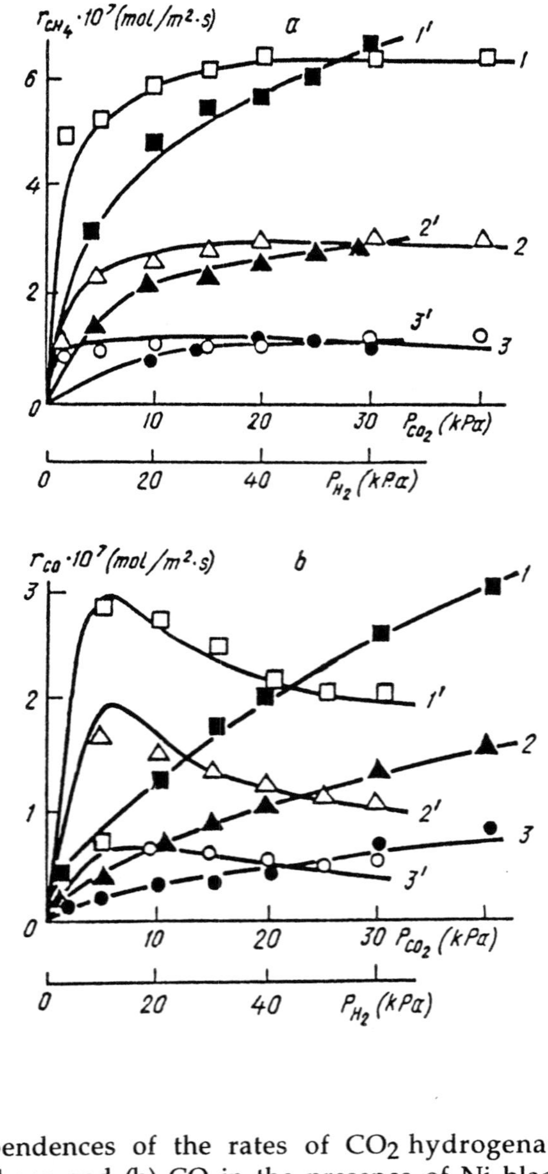

Figure 13 Dependences of the rates of CO_2 hydrogenation to (a) methane and (b) CO in the presence of Ni black on (1,2,3) P_{CO_2} and (1',2',3') P_{H_2} at (1,1') 574°K, (2,2') 553°K, and (3,3') 531°K.

$$r = k_3 \frac{b_{CO_2} P_{CO_2} \sqrt{b_{H_2} P_{H_2}}}{\left(1 + b_{CO_2} P_{CO_2} + \sqrt{b_{H_2} P_{H_2}}\right)^2} \tag{47}$$

If step (4) is the limiting one while step (3) is in equilibrium, the kinetic equation is

$$r = k_4 \frac{K_3 b_{CO_2} P_{CO_2} b_{H_2} P_{H_2}}{\left[1 + b_{CO_2} P_{CO_2} + \sqrt{b_{H_2} P_{H_2}}\left(1 + K_3 b_{CO_2} P_{CO_2}\right)\right]^2} \tag{48}$$

A version of scheme (XXXII) is the mechanism where equilibrium adsorption of the reagents takes place at different active sites of the catalyst surface (Z_C and Z_H):

(1) $H_2 + 2Z_H = 2Z_HH$

(2) $CO_2 + Z_C = Z_CCO_2$

(3) $Z_CCO_2 + Z_HH \rightarrow Z_CCO_2H + Z_H$

(4) $Z_CCO_2H + Z_HH \rightarrow Z_CC(OH)_2 + Z_H$

(5) $Z_CC(OH)_2 + Z_HH \rightarrow Z_CCOH + H_2O + Z$

(6) $Z_CCOH + Z_HH \rightarrow Z_CCHOH \xrightarrow{Z_HH} \ldots CH_4 + H_2O + Z_C + Z_H$

(XXXIII)

With mechanism (XXXIII) we obtain the following kinetic equations:

--for limiting step (3)

$$r = k_3 \frac{b_{CO_2} P_{CO_2}}{1 + b_{CO_2} P_{CO_2}} \cdot \frac{\sqrt{b_{H_2} P_{H_2}}}{1 + \sqrt{b_{H_2} P_{H_2}}} \tag{49}$$

for limiting step (4)

$$r = k_4 \frac{K_3 b_{CO_2} P_{CO_2} \sqrt{b_{H_2} P_{H_2}}}{1 + b_{CO_2} P_{CO_2} + K_3 b_{CO_2} P_{CO_2} \sqrt{b_{H_2} P_{H_2}}} \cdot \frac{\sqrt{b_{H_2} P_{H_2}}}{1 + \sqrt{b_{H_2} P_{H_2}}} \tag{50}$$

Taking account of the data available in the literature on the lack of CO_2 adsorption on transition metals [57], we have considered different versions of a CO_2-impact mechanism of methane synthesis (reaction IX):

$$\begin{aligned}
&(1)\ H_2 + 2Z = 2ZH \\
&(2)\ CO_2 + ZH \rightarrow ZCO_2H \\
&(3)\ ZCO_2H + ZH \rightarrow ZC(OH)_2 + Z \\
&(4)\ ZC(OH)_2 + ZH \rightarrow ZCOH + H_2O + Z \\
&(5)\ ZCOH + ZH \rightarrow ZCHOH \xrightarrow{ZH} \ldots CH_4 + H_2O + Z
\end{aligned}$$

(XXXIV)

On condition that the limiting step is the interaction of a CO_2 molecule from the gas phase with ZH (step 2), the corresponding kinetic equation is

$$r = k_2 \frac{P_{CO_2}\sqrt{b_{H_2}P_{H_2}}}{1+\sqrt{b_{H_2}P_{H_2}}} \tag{51}$$

If step (2) is in equilibrium whereas step (3) is limiting, the kinetic equation for mechanism (XXXIV) becomes

$$r = k_3 \frac{K_2 P_{CO_2} b_{H_2} P_{H_2}}{\left[1+\sqrt{b_{H_2}P_{H_2}}\left(1+K_2P_{CO_2}\right)\right]^2} \tag{52}$$

A version of scheme (XXXIV) is the mechanism where in the limiting step (2) a CO_2 molecule from the gas phase interacts with two adsorbed hydrogen atoms:

$$\begin{aligned}
&(1)\ H_2 + 2Z = 2ZH \\
&(2)\ CO_2 + 2ZH \rightarrow Z_2C(OH)_2 \\
&(3)\ Z_2C(OH)_2 + ZH \rightarrow ZCOH + H_2O + 2Z \\
&(4)\ ZCOH + ZH \rightarrow ZCHOH \xrightarrow{ZH} \ldots CH_4 + H_2O + Z
\end{aligned}$$

(XXXV)

with a corresponding kinetic equation

$$r = k_2 P_{CO_2} \cdot \left(\frac{\sqrt{b_{H_2} P_{H_2}}}{1 + \sqrt{b_{H_2} P_{H_2}}} \right)^2 \tag{53}$$

If step (2) in this scheme is in equilibrium and the third step is limiting, the rate of methane formation would be expressed as

$$r = k_3 \frac{\left[\sqrt{\left(1 + \sqrt{b_{H_2} P_{H_2}}\right)^2 + 4 K_2 P_{CO_2} b_{H_2} P_{H_2}} - 1 - \sqrt{b_{H_2} P_{H_2}} \right]^3}{8 \left(K_2 P_{CO_2}\right)^2 \cdot \left(b_{H_2} P_{H_2}\right)^{3/2}} \tag{54}$$

Computer analysis of equations (47) to (54) reveals that the observed kinetic dependences of CO_2 reduction to methane are adequately described by equation (54); this is illustrated in Figure 12 where the points are the experimental data and the curves have been generated with equation (54).

Further, mechanism (XXXV) with the second step in equilibrium was taken as the basis for constructing the complete mechanism of CO_2 hydrogenation to methane and carbon monoxide. Computer analysis of about 80 different versions has shown that the whole body of experimental data can be described using the following mechanism:

(1) $H_2 + 2Z = 2ZH$

(2) $CO_2 + 2ZH = Z_2C(OH)_2$

(3) $CO_2 + ZH \rightarrow ZCO_2H$

(4) $ZCO_2H + ZH \rightarrow ZCO + ZH_2O$

(5) $Z_2C(OH)_2 + ZH = Z_2HC(OH)_2 + Z$

(6) $Z_2HC(OH)_2 + ZH = Z_2H_2C(OH)_2 + Z$

$$(7)\ Z_2H_2C(OH_2) \rightarrow ZCHOH + ZH_2O$$
$$(8)\ ZCO_2H \rightarrow CO + ZOH$$
$$(9)\ ZCO + ZH \rightarrow ZCOH + Z$$
$$(10)\ ZCOH + ZH = ZCHOH + Z$$
$$(11)\ ZCHOH + ZH \rightarrow ZCH_2 + ZH_2O$$
$$(12)\ ZCH + ZH \rightarrow ZCH_2 + Z$$
$$(13)\ ZCH_2 + ZH \rightarrow ZCH_3 + Z$$
$$(14)\ ZCH_3 + ZH \rightarrow ZCH_4 + Z$$
$$(15)\ ZCH_4 \rightarrow CH_4 + Z$$
$$(16)\ ZH_2O \rightarrow H_2O + Z$$

(XXXVI)

The corresponding overall scheme for carbon dioxide conversion can be presented as follows:

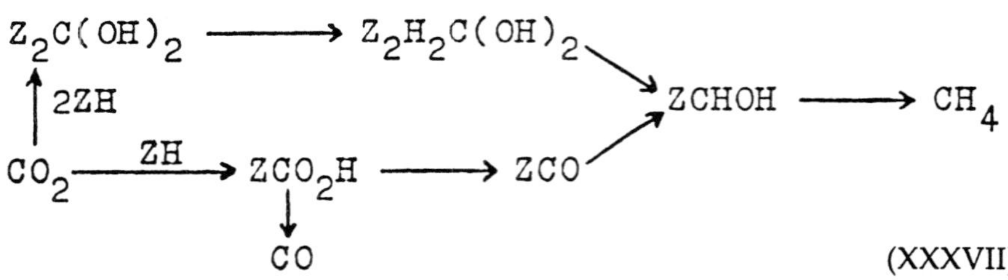

(XXXVII)

According to mechanism (XXXVI) and chart (XXXVII), CO_2 interacts with two hydrogen atoms, and then hydrogens are consecutively added to the adsorbed complex $Z_2C(OH)_2$ in several quasi-equilibrium steps to form the intermediate $Z_2H_2C(OH)_2$, decomposition of which in step (7) determines the rate of methane formation from CO_2. If in step (3) CO_2 reacts with only one hydrogen atom, a ZCO_2H complex is formed which either decomposes to yield CO or is hydrogenated to CH_4. Mechanism (XXXVI) is reflected in the following set of kinetic equations for the rates of methane and CO formation:

$$r_{CH_4} = k_7K_6K_5K_2P_{CO_2}\left(b_{H_2}P_{H_2}\right)^2\left(\Theta^o\right)^2 + k_4\frac{k_3P_{CO_2}b_{H_2}P_{H_2}\left(\Theta^o\right)^2}{k_4\Theta^o\sqrt{b_{H_2}P_{H_2}} + k_8} \tag{55}$$

$$r_{CO} = k_8 \frac{k_3 P_{CO_2} \sqrt{b_{H_2} P_{H_2}} \cdot \Theta^o}{k_4 \Theta^o \sqrt{b_{H_2} P_{H_2}} + k_8} \tag{56}$$

where Θ^O is obtained by solving the balance equation

$$a\left(\Theta^o\right)^3 + b\left(\Theta^o\right)^2 + c\left(\Theta^o\right) + d = O \tag{57}$$

with

$$a = k_4 k_2 P_{CO_2} \left(b_{H_2} P_{H_2}\right)^{3/2} \left[1 + K_5 \sqrt{b_{H_2} P_{H_2}} \left(1 + K_6 \sqrt{b_{H_2} P_{H_2}}\right)\right]$$

$$b = k_8 K_2 P_{CO_2} b_{H_2} P_{H_2} \left[1 + K_5 \sqrt{b_{H_2} P_{H_2}} \left(1 + K_6 \sqrt{b_{H_2} P_{H_2}}\right)\right] +$$

$$+ k_4 \sqrt{b_{H_2} P_{H_2}} \left(1 + \sqrt{b_{H_2} P_{H_2}}\right)$$

$$c = k_8 \left(1 + \sqrt{b_{H_2} P_{H_2}}\right) + k_3 P_{CO_2} \sqrt{b_{H_2} P_{H_2}} - k_4 \sqrt{b_{H_2} P_{H_2}}$$

$$d = -k_8$$

The set of equations (55)–(57) provides a quantitative description of the whole body of experimental data that we have obtained on the kinetics of hydrogenation of CO_2 to methane and CO on transition metals; this is exemplified in Figure 13 where the curves have been computed and the points obtained experimentally.

The kinetic and thermodynamic constants calculated using this model conform to their physical meaning: the rate constants increase while the adsorption coefficients decrease with temperature.

It is noteworthy that during hydrogenation of CO_2, unlike that of CO, hydrocarbons higher than C_2 are either not formed at all (for instance, on Rh) or appear in minor amounts. Our comparison of the rates of hydrocarbon formation from CO and CO_2 [92] allows a conclusion that methane homologs do not arise directly from CO_2 but only through intermediate formation of CO.

6. CONCLUSION

Some general conclusions can be made on the strength of the data presented in this review on the kinetics and mechanisms of hydrogen addition to compounds of various classes. The peculiarity of processes involving molecular hydrogen is the great diversity of their kinetic models. Such models are often quite different for similar processes, like hydrogenation of carbonyl compounds and carbon monoxide, or partial and complete reduction of the C = O group. This is due to the wide spectrum of possible intermediate adsorbed forms of the reacting molecules and the strength of their binding to the catalyst.

The kinetics of the process is substantially complicated when the hydrogenation reaction is accompanied by acid–base hydrocondensation conversions of the initial compounds and intermediates, as it happens in the synthesis of secondary and tertiary amines from nitriles, or synthesis of higher hydrocarbons.

A complete kinetic description of the hydrogenation reaction and similar processes requires sufficient information on their mechanism obtained by physicochemical techniques.

It is to be hoped that the general kinetic patterns and peculiarities of the reactions involving molecular hydrogen presented in this review will be helpful in further generalization of new experimental results concerning specific reactions, and will serve a basis for studying the mechanism of hydrogenation reactions, simulation and optimization of these processes.

Acknowledgement

The authors wish to express their gratitude to Professor G.I. Golodets, Corresponding Member of the Ukrainian Academy of Sciences, for his constant attention to their work, valuable suggestions and discussions.

7. References

1. **M. I. Temkin**, In: "Mekhanizm i kinetika slozhnykh reaktsiĭ" (Mechanism and Kinetics of Complex Reactions), Moscow: Nauka, 1970, p. 57, Russian.
2. **N. V. Pavlenko**, Kataliz i Katalizatory, vol. 21, 1983, p. 55, Russian.
3. **E. V. Prokhorenko, N. V. Pavlenko**, Ukrainskiĭ Khimicheskiĭ Zhurnal, vol. 53, no. 7, 1987, p. 785, Russian.
4. **N. V. Pavlenko, A. T. Tripolskiĭ, G. I. Golodets**, Kinetika i Kataliz, vol. 30, no. 6, p. 1371, Russian.
5. **D. Stall, E. Westram, G. Zinke**, The Chemical Thermodynamics of Organic Compounds, New York: John Wiley and Sons, Inc., 1969, 807 pp.
6. **Yu. M. Zhorov**, Termodinamika khimicheskikh protsessov (Thermodynamics of Chemical Processes), Moscow: Khimiya, 1985, 458 pp., Russian.
7. **G. T. H. Stoddart, G. Kemball**, J. Coll. Sci., vol. 11, 1956, p. 532.
8. **S. Watanabe**, Sci. and Ind., vol. 31, 1957, p. 195.

9. **S. L. Kiperman, G. I. Kaplan**, Kinetika i Kataliz, vol. 5, no. 5, 1964, p. 888, Russian.
10. **A. M. Sokolskaya**, Kataliticheskie reaktsii v zhidkoĭ faze (Catalytic Reactions in Liquid Phase), Alma-Ata: Nauka, 1972, 317 pp., Russian.
11. **G. Herbo**, J. Chim. Phys. et Phys.-Chim. Biol., vol. 47, no. 8, 1950, p. 454.
12. **T. Freund, H. M. Hulburt**, J. Phys. Chem., vol. 61, no. 7, 1957, p. 909.
13. **T. P. Gegenava**, Izvestiya AN GruzSSR, Ser. Khim., vol. 2, no. 3, 1976, p. 259, Russian.
14. **P. B. Babkova, A. K. Avetisov, G. D. Lyubarskiĭ et al.**, Kinetika i Kataliz, vol. 10, no. 5, 1969, p. 1086, Russian.
15. **P. B. Babkova, A. K. Avetisov, G. D. Lyubarskiĭ et al.**, Kinetika i Kataliz, vol. 13, no. 4, 1972, p. 915, Russian.
16. **P. B. Babkova, A. K. Avetisov, G. D. Lyubarskiĭ et al.**, Kinetika i Kataliz, vol. 11, no. 6, 1970, p. 1451, Russian.
17. **P. B. Babkova, A. K. Avetisov, G. D. Lyubarskiĭ et al.**, Kinetika i Kataliz, vol. 13, no. 2, 1972, p. 345, Russian.
18. **T. Kwan**, Kataliz. Issledovanie geterogennykh protsessov (Catalysis. Investigation of Heterogeneous Processes) Moscow: IL, 1956, p. 314, Russian.
19. **V. F. Khorkov, G. D. Lyubarskiĭ, A. I. Gelbshteĭn**, Zhurnal Fizicheskoĭ Khimii, vol. 46, no. 11, p. 2853, Russian.
20. **G. H. Twigg**, Disc. Farad. Soc., vol. 8, 1950, p. 152.
21. **L. O. Apelbaum, M. I. Temkin**, Zhurnal Fizicheskoĭ Khimii, vol. 35, no. 9, 1961, p. 2060.
22. **Ch. Satterfield**, Heterogeneous Catalysis in Practice, New York: McGraw–Hill, 1980, 520 pp.
23. **S. L. Kiperman**, Uspekhi Khimii, vol. 47, no. 1, 1978, p. 3, Russian.
24. **J. Simonicova, L. Hillaire, J. Panek et al.**, Z. Phys. Chem. Neue Folge. Bd., vol. 83, 1973, p. 287.
25. **N. V. Pavlenko, A. I. Tripolskiĭ, G. I. Golodets et al.**, Kinetika i Kataliz, vol. 26, no. 1, 1985, p. 115, Russian.
26. **A. I. Tripolskiĭ, N. V. Pavlenko, G. I. Golodets**, Kinetika i Kataliz, vol. 26, no. 5, 1985, p. 1131, Russian.
27. **N. V. Pavlenko, A. I. Tripolskiĭ, G. I. Golodets**, Theor. Exptl. Chem., vol. 22, no. 6, 1986, p. 667.
28. **A. I. Tripolskiĭ, N. V. Pavlenko, G. M. Tebliz, G. I. Golodets**, React. Kinet. Catal. Lett., vol. 22, no. 3–4, 1983, p. 367.
29. **N. V. Pavlenko, A. I. Tripolskiĭ, G. I. Golodets et al.**, Kinetika i Kataliz, vol. 26, no. 1, 1985, p. 108, Russian.
30. **G. I. Golodets, N. V. Pavlenko, A. I. Tripolskiĭ et al.**, Kinetika i Kataliz, vol. 26, no. 3, 1985, p. 637, Russian.

31. **A. I. Tripolskiĭ, N. V. Pavlenko, G. I. Golodets**, Kinetika i Kataliz, vol. 26, no. 6, 1985, p. 1494, Russian.
32. **G. I. Golodets, N. V. Pavlenko, A. I. Tripolskiĭ**, Kinetika i Kataliz, vol. 27, no. 2, 1986, p. 346, Russian.
33. **N. V. Pavlenko, A. I. Tripolskiĭ, G. I. Golodets**, In: IVth All-Union Conference on the Mechanisms of Catalytic Reactions, Moscow 1986, Preprint no. 84, p. 385, Russian.
34. **N. V. Pavlenko, A. I. Tripolskiĭ, G. I. Golodets**, Kinetika i Kataliz, vol. 28, no. 2, 1987, p. 437, Russian.
35. **N. V. Pavlenko, A. I. Tripolskiĭ, G. I. Golodets**, Teoreticheskaya i Eksperimentalnaya Khimiya, vol. 23, no. 2, 1987, p. 250, Russian.
36. **G. I. Golodets**, Heterogeneous Catalytic Reactions Involving Molecular Oxygen, New York: Elsevier, 1983, 878 pp.
37. **N. V. Pavlenko, A. I. Tripolskiĭ, G. I. Golodets**, Kinetika i Kataliz, vol. 30, no. 6, 1989, p. 1371, Russian.
38. **A. I. Subbotin, I. V. Nekrasov, V. I. Yakerson et al.**, Kinetika i Kataliz, vol. 29, no. 4, 1988, p. 840, Russian.
39. **A. I. Subbotin, B. S. Gudkov, N. V. Nekrasov et al.**, Kinetika i Kataliz, vol. 29, no. 5, 1988, p. 1136, Russian.
40. **A. I. Subbotin, B. S. Gudkov, N. V. Nekrasov et al.**, Kinetika i Kataliz, vol. 29, no. 5, 1988, p. 1141, Russian.
41. **L. S. Glebov, G. A. Kliger, A. N. Bashkirov**, Neftekhimiya, vol. 18, no. 5, 1978, p. 803, Russian.
42. **L. S. Glebov, G. A. Kliger, A. N. Bashkirov et al.**, Kinetika i Kataliz, vol. 19, no. 5, 1978, p. 1339, Russian.
43. **L. Kh. Freĭdlin, T. A. Sladkova**, Uspekhi Khimii, vol. 33, no. 6, 1964, p. 664, Russian
44. **M. V. Klyuev, O. P. Savateev, Yu. G. Erykalov**, In: "Voprosy kinetiki i kataliza. Zakonomernosti formirovaniya geterogennykh katalizatorov" (Problems of Kinetics and Catalysis. The Pattern of Formation of Heterogeneous Catalysts), Ivanovo: Vysshaya Shkola, 1983, p. 67, Russian.
45. **N. V. Pavlenko, A. I. Tripolskiĭ, E. V. Prokhorenko et al.**, Kinetika i Kataliz, vol. 28, no. 6, 1987, p. 1382, Russian.
46. **N. V. Pavlenko, A. I. Tripolskiĭ, E. V. Prokhorenko et al.**, Teoreticheskaya i Eksperimentalnaya Khimiya, vol. 24, no. 5, 1988, p. 546, Russian.
47. **N. V. Pavlenko, A. I. Tripolskiĭ, O. T. Popovich et al.**, Kinetika i Kataliz, vol. 30, no. 4, 1989, p. 848, Russian.
48. **E. N. Zilberman**, Reaktsii nitrilov (Reactions of Nitriles), Moscow: Khimiya, 1972, 446 pp., Russian.
49. **S. L. Kiperman**, Osnovy khimicheskoĭ kinetiki v geterogennom katalize (Principles of Chemical Kinetics in Heterogeneous Catalysis), Moscow: Khimiya, 1979, 386 pp., Russian.

50. **G. I. Golodets, N. V. Pavlenko, E. V. Prokhorenko**, Kinetika i Kataliz, vol. 28, no. 3, 1987, p. 625, Russian.
51. **N. V. Pavlenko, A. I. Tripolskiĭ, G. I. Golodets**, Kinetika i Kataliz, vol. 29, no. 3, 1988, p. 746, Russian.
52. **Yu. A. Zhdanov, V. I. Minkin**, Korrelyatsionnyĭ analiz v organicheskoĭ khimii (Correlation Analysis in Organic Chemistry), Rostov: Rostovskiĭ Universitet, 1966, 469 pp., Russian.
53. **M. A. Vannice**, Catal. Rev. Sci. Eng., vol. 14, no. 2, 1976, p. 153.
54. **F. Hiroshi, I. Nobuhide, M. Isao**, J. Catal., vol. 100, no. 2, 1986, p. 279.
55. **R. E. Hayes, W. J. Thomas, K. E. Hayes**, J. Catal., vol. 92, no. 2, 1985, p. 312.
56. **S. C. Kellner, A. T. Bell**, J. Catal., vol. 70, no. 2, 1981, p. 418.
57. **V. M. Vlasenko, G. E. Yuzefovich**, Uspekhi Khimii, vol. 38, no. 9, 1969, p. 1622, Russian.
58. **J. R. H. Ross**, J. Catal., vol. 71, no. 1, 1981, p. 205, Russian.
59. **T. Herwijnen, H. Doesburg, W. A. Jang**, J. Catal., vol. 28, no. 3, 1973, p. 391.
60. **M. A. Vannice**, J. Catal., vol. 44, no. 1, 1976, p. 152.
61. **C. P. Huang, J. T. Richardson**, J. Catal., vol. 51, no. 1, 1978, p. 1.
62. **S. V. Ho, P. Harriott**, J. Catal., vol. 64, no. 2, 1987, p. 272.
63. **J. Klose, M. Baerns**, J. Catal., vol. 85, no. 1, 1984, p. 105.
64. **R. Z. C. Meerten, J. G. Vellenbrock, M. H. J. Croon et al.**, Appl. Catal., vol. 3, no. 1, 1982, p. 29.
65. **Z. A. Ibraeva, N. V. Nekrasov, B. S. Gudkov, V. I. Yakerson, E. Z. Golosman, Z. T. Beĭsembaeva, S. L. Kiperman**, In: "Khimicheskaya kinetika v katalize. Kineticheskie modeli neorganicheskikh reaktsiĭ" (Chemical Kinetics in Catalysis. Kinetic Models of Inorganic Reactions), Chernogolovka, 1985, p. 47, Russian.
66. **Z. A. Ibraeva, N. V. Nekrasov, B. S. Gudkov, V. I. Yakerson, E. Z. Golosman, Z. T. Beĭsembaeva, S. L. Kiperman**, Kinetika i Kataliz, vol. 28, no. 2, 1987, p. 386, Russian.
67. **G. I. Golodets, N. V. Pavlenko, A. I. Tripolskiĭ**, React. Kinet. Catal. Lett., vol. 32, no. 2, 1986, p. 481.
68. **G. I. Golodets, N. V. Pavlenko, A. I. Tripolskiĭ**, Kinetika i Kataliz, vol. 28, no. 4, 1987, p. 883, Russian.
69. **G. I. Golodets, N. V. Pavlenko, A. I. Tripolskiĭ**, Kinetika i Kataliz, vol. 28, no. 4, 1987, p. 888, Russian.
70. **G. I. Golodets, N. V. Pavlenko, A. I. Tripolskiĭ**, React. Kinet. Catal. Lett., vol. 34, no. 2, 1987, p. 373.
71. **N. V. Pavlenko, E. V. Prokhorenko, A. I. Tripolskiĭ et al.**, Kinetika i Kataliz, vol. 30, no. 6, 1989, p. 1364, Russian.
72. **G. I. Golodets, N. V. Pavlenko, E. V. Prokhorenko et al.**, React. Kinet. Catal. Lett., vol. 38, no. 1, 1989, p. 119.

73. **N. V. Pavlenko**, Uspekhi Khimii, vol. 58, no. 5, 1989, p. 767, Russian.
74. **N. V. Pavlenko, E. V. Prokhorenko, G. M. Telbiz et al.**, Teoreticheskaya i Eksperimentalnaya Khimiya, vol. 25, no. 6, 1989, p. 693, Russian.
75. **G. I. Golodets**, Teoreticheskaya i Eksperimentalnaya Khimiya, vol. 21, no. 5, 1985, p. 525, Russian.
76. **M. A. Vannice**, J. Catal., vol. 37, no. 3, 1975, p. 449.
77. **Yu. B. Naumov, E. S. Startseva, Yu. I. Berezina et al.**, Kinetika i Kataliz, vol. 25, no. 6, 1984, p. 1365, Russian.
78. **V. M. Vlasenko, M. T. Rusov, G. E. Yuzefovich**, Kinetika i Kataliz, vol. 2, no. 4, 1961, p. 525, Russian.
79. **J. H. Chlang, J. R. Hopper**, Ind. Eng. Chem. Prod. Res. Develop., vol. 22, no. 2, 1983, p. 225.
80. **J. Muller, V. Pour, A. Regnier**, J. Catal., vol. 11, no. 4, 1968, p. 326.
81. **F. Goodringe, H. A. Quazi**, Trans. Inst. Chem. Eng., vol. 45, no. 2, 1967, p. 274.
82. **M. Sols**, Coll. Czech. Chem. Commun., vol. 27, no. 2, 1962, p. 2621.
83. **F. Solymosi, A. Erdohelyi, T. Bansagi**, J. Catal., vol. 68, no. 2, 1981, p. 371.
84. **F. Solymosi, A. Erdohelyi, M. Kocsis,** J. Chem. Soc. Faraday Trans. I, vol. 77, no. 5, 1981, p. 1003.
85. **V. A. Naumov, O. V. Krylov**, Kinetika i Kataliz, vol. 20, no. 5, 1979, p. 1344, Russian.
86. **V. A. Naumov, O. V. Krylov**, Kinetika i Kataliz, vol. 20, no. 5, 1979, p. 1347, Russian.
87. **R. Maatman, S. Hiemstra**, J. Catal., vol. 62, no. 2, 1980, p. 349.
88. **F. Mey, H. Spinder, W. Rolander, H. Wejbenborn**, Chem. Techn., (DDR). vol. 38, no. 7, 1986, p. 300.
89. **G. D. Weatherbee, C. H. Bartholomew**, J. Catal., vol. 77, no. 2, 1982, p. 460.
90. **G. D. Zakumbaeva, Sh. U. Urumbaeva, D. G. Nigmetova et al.**, Izvestiya AN KazSSR, Ser. Khim., no. 3, 1985, p. 19, Russian.
91. **Yu. M. Serov, V. M. Zhernosek, V. M. Gryaznov**, Kinetika i Kataliz, vol. 24, no. 1, 1983, p. 246; no. 2, p. 362, Russian.
92. **A. I. Tripolskiĭ, N. V. Pavlenko**, In: All-Union Symposium on the Problem of Kinetics in Carbon Chemistry, 1990, Preprint no. 42, p. 104, Russian.
93. **R. Barget, J. Trambouze**, C. R. Acad. Sci., vol. 290, no. 9, 1980, p. 153.
94. **H. Q. Dao, F. Pruchnik**, Bull. Pol. Pol. Acad. Sci. Chem., vol. 32, no. 12, 1984, p. 475.
95. **F. Solymosi, A. Erdohelyi, M. Kocsis**, J. Catal., vol. 65, no. 3, 1980, p. 428.

KINETICS OF THE INITIAL STEP OF HYDROCARBON OXIDATION CATALYZED BY ONIUM SALTS

I.A. Opeĭda, N.M. Zalevskaya
Institute of Physico-Organic Chemistry and Coal Chemistry,
Ukrainian Academy of Sciences,
Donetsk, Soviet Union

Abstract—The kinetic dependences of oxygen consumption at the initial step of cumene oxidation to a hydroperoxide in the presence of n-butyl pyridinium bromide have been used for formal kinetic analysis and simulation of the kinetics of the chain processes of hydrocarbon oxidation in which initiation involves the end product of the reaction. A number of kinetic parameters have been obtained for these processes. It has been shown that an adequate description of the reaction under study (cumene oxidation) should account not only for the steps of initiation and molecular decomposition of the hydroperoxide in its reaction with the catalyst. The efficiency of n-butyl pyridinium bromide has been found to increase with a decrease of its concentration. The applicability of the classical scheme of oxidation and the criteria used to assess its validity have been considered.

1. INTRODUCTION

Our previous works [1–5] have been concerned with the action of onium salts in hydrocarbon oxidation. The experimental data obtained [1–5] point to that the main route of initiation of this process is the decomposition of hydroperoxides under the action of the catalyst. The observed kinetic dependences were interpreted using the concepts of the mechanism of catalysis developed quite some time ago [6–7] and applied in all subsequent studies. However, the kinetic information obtained in our works has revealed some peculiarities of the process that require further investigation. The present work is aimed at a more detailed analysis of the hydrocarbon oxidation mechanism for the case when the process is initiated at the expense of hydroperoxide decomposition under the action of onium salts.

2. EXPERIMENTAL

Cumene was used as a model compound. Cumene and the solvents were purified according to established procedures [8]. The catalyst was n-butyl pyridinium bromide; it was purified by precipitating the salt with carbon tetrachloride from acetonitrile, collecting the precipitate on a filter and drying it *in vacuo*.

The methods of measuring the oxygen consumption rate and the content of hydroperoxide groups were described in our earlier work [2]. The determination accuracy was within 5% for consumed oxygen and within 6% for hydroperoxide concentration.

The set of differential equations was solved according to Runge–Kutt, minimizing the sum of square deviation of the calculated concentration values from the experimentally determined ones. Data processing was performed for 17 kinetic curves of oxygen consumption and hydroperoxide accumulation during cumene oxidation in benzonitrile at 343K (without prior introduction of hydroperoxide, when $C_{1o} \leq 0.001$ mole/l; and with such introduction, when $C_{1o} = 0.326$ mole/l) and at 358 K. The analysis of the kinetic curves was restricted to the time interval from the onset of reaction till the moment when the rate started to decline after reaching W_m, so that the extent of the oxidative conversion of cumene was 3 to 6% (Figure 1). At higher extents of conversion (about 10%) the description may become less accurate because of substantial consumption of cumene, which would require complication of the model. Cumyl hydroperoxide concentrations were calculated using an empirical dependence [5]

$$C_1 = C_{10} + 0.795\, C_3 - 0.392\, C_3^2 \text{ (mole/l).}$$

3. PATTERNS OF CATALYZED OXIDATION OF HYDROCARBONS TO HYDROPEROXIDES

In a general form, the scheme of catalyzed oxidation of hydrocarbons to hydroperoxides includes the following steps [6]:

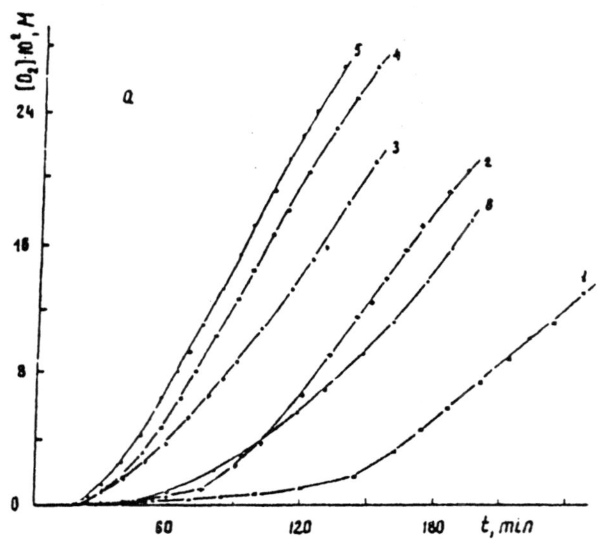

Figure 1a

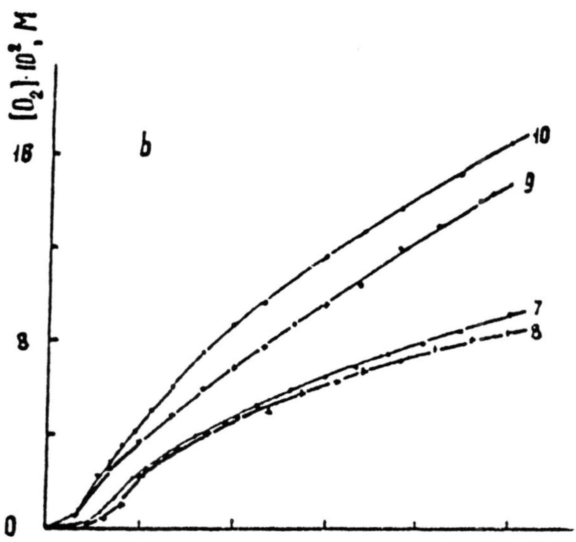

Figure 1b

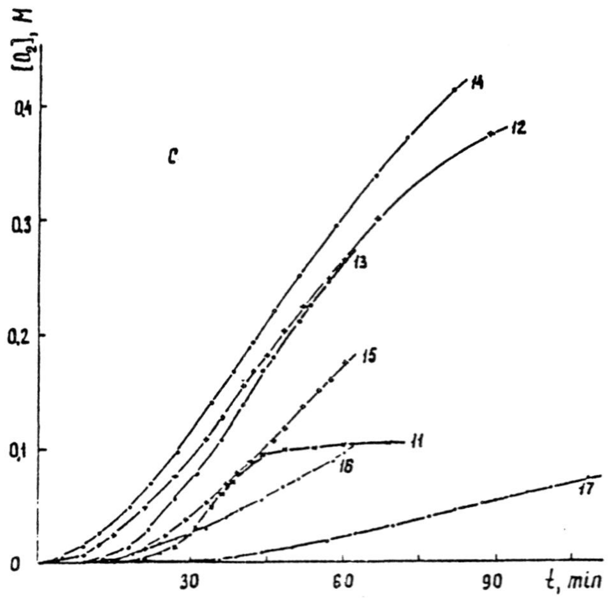

Figure 1. Kinetic curves of oxygen consumption during cumene oxidation in benzonitrile.

(*a*) 343K; [RH] = 3.58 mole/l; $C_{20} \cdot 10^3$ (mole/l): (1) 0.17, (2) 0.35, (3) 3.5, (4) 9.0, (5) 12.3, (6) 25.7;
(*b*) 343K; [RH] = 3.58 mole/l, C_{10} = 0.326 mole/l; $C_{20} \cdot 10^3$ (mole/l): (7) 0.029, (8) 0.058, (9) 0.26, (10) 4.6;
(*c*) 358K; [RH] (mole/l): (11–14) 3.58, (15) 2.87, (16) 2.15, (17) 1.43; $C_{20} \cdot 10^3$ (mole/l): (11) 0.07, (12) 2.2, (14) 12.0, (13, 15–17) 4.5.

1. Chain initiation involving a catalyst;
2. Chain propagation:

$$R^\bullet + O_2 \rightarrow ROO^\bullet, ROO^\bullet + RH \xrightarrow{k_p} ROOH + R^\bullet;$$

3. Chain termination:

$$2ROO^\bullet \xrightarrow{k_t} \text{inactive products } + O_2,$$

where ROOH is the hydroperoxide, $R^\cdot$ and $ROO^\cdot$ are the alkyl and the peroxide radicals.

In our earlier works [3,4] on cumene and acenaphthene oxidation we have observed that the dependence of the maximal oxidation rate (W_m) on the square of hydrocarbon concentration ([RH]) yields chain propagation (k_p) and chain termination (k_t) rate constants whose ratios $\left(k_p/\sqrt{k_t}\right)$ differs approximately threefold from those measured under conditions of initiated oxidation. Conventionally this phenomenon is explained by the contribution of the reactions of nonradical decomposition of the hydroperoxide [7], and the maximal rate is expressed as

$$W_m = \frac{k_p^2[RH]^2}{k_t} \cdot \frac{K_i}{K_i + K_m} = C\frac{k_p^2[RH]^2}{k_t}, \qquad C = \frac{K_i}{K_i + K_m}, \tag{1}$$

where K_i and K_m are the rate constants for the radical and the molecular decomposition of hydroperoxide. If one takes into account that the $k_p/\sqrt{k_t}$ value is independent of the oxidation mechanism, then for cumene and acenaphthene oxidation in the presence of n-butyl pyridinium bromide the coefficient C must by equal to 0.11, reflecting the portion of hydroperoxide decomposing via the radical route. On the other hand, within the framework of the mechanism excluding a nonradical way of hydroperoxide consumption [6] the coefficient has another meaning and may be equal to 1 or 4 depending on the initiation mechanism.

Analysis of oxidation catalyzed by onium salts also reveals another substantial discrepancy between the supposed mechanism and the observed kinetic patterns. The classical mechanism predicts a maximal rate in the region of the steady-state (which in this case is also the maximal) hydroperoxide concentration. Yet in our experiments where the concentrations have been widely varied both for the catalyst and for the hydrocarbon and hydroperoxide [4,5] the rate of oxygen consumption passes through a maximum at hydroperoxide concentrations that are lower by far than the steady-state ones.

4. Initiation Mechanism

One of the key questions here is the one of the initiation mechanism or, more specifically, of oxygen consumption at the first step. Since chain initiation mainly takes place through catalytic decomposition of the hydroperoxide [6] the following variants are possible.

(i) Decomposition involving one molecule of hydroperoxide:

$$\mathrm{Ct + ROOH \rightarrow RO^{\bullet} + {}^{\bullet}OH} \qquad (k_i)$$
$$\mathrm{{}^{\bullet}OH + RH \rightarrow R^{\bullet} + H_2O}$$
$$\mathrm{RO^{\bullet} + RH \rightarrow R^{\bullet} + ROH}$$

where Ct is the catalyst, ROH is the alcohol, $RO^{\bullet}$ and $^{\bullet}OH$ are the alkoxyl and the hydroxyl radicals, respectively, and k_i is the initiation rate constant. With such initiation, each alkyl radical formed reacts with a molecule of oxygen, therefore the oxygen consumption rate at this step is $2k_i$[ROOH][Ct].

(ii) Decomposition involving two hydroperoxide molecules:

$$\mathrm{Ct + ROOH \rightarrow ROO^{\bullet} + H^+}$$
$$\mathrm{Ct + ROOH \rightarrow RO^{\bullet} + OH^-}$$
$$\mathrm{RO^{\bullet} + HR \rightarrow R^{\bullet} + ROH}$$

Here the overall reaction can be written down as

$$\mathrm{2ROOH + RH \rightarrow ROO^{\bullet} + R^{\bullet} + ROH + H_2O}$$

and the oxygen consumption rate at the initiation step is in this case equal to k_i[ROOH][Ct].

(iii) Initiation without oxygen consumption:

(a)

$$\mathrm{Ct + ROOH \rightarrow RO^{\bullet} + OH^-}$$
$$\mathrm{Ct + ROOH \rightarrow ROO^{\bullet} + H^+}$$
$$\mathrm{RO^{\bullet} + ROOH \rightarrow ROH + ROO^{\bullet}}$$

or (b)

$$\mathrm{Ct + ROOH \rightarrow RO^{\bullet} + {}^{\bullet}OH}$$
$$\mathrm{RO^{\bullet} + ROOH \rightarrow ROO^{\bullet} + ROH}$$
$$\mathrm{{}^{\bullet}OH + ROOH \rightarrow ROO^{\bullet} + H_2O}$$

The overall reaction of hydroperoxide decomposition in this case can be written down as

$$\mathrm{3ROOH \rightarrow 2ROO^{\bullet} + ROH + H_2O.}$$

Thus, chain initiation in the presence of a catalyst can proceed via different mechanisms whereby decomposition of a single hydroperoxide molecule gives rise to m radicals, with $0 < m \leq 2$. In the general form this can be put down as

$$Ct + l\text{ROOH} \rightarrow s\text{R}^{\cdot} + f\text{ROO}^{\cdot} \tag{2}$$

with $m = (s + f)/l$ where $s, f,$ and l are the corresponding stoichiometric coefficients.

Under steady-state conditions, the initiation rate is equal to the termination rate. Taking this into account and assuming the reaction of the catalyst with one hydroperoxide molecule as the rate-limiting step of initiation, we can write down the equation

$$amx = k_t[\text{ROO}^{\cdot}]^2 \tag{3}$$

where $a = k_i[\text{Ct}]_0$ (ignoring the consumption of catalyst for simplicity), and x = [ROOH] is the current concentration of hydroperoxide.

Analysis of schemes (i) to (iii) outlined above shows that the number of oxygen molecules consumed (n) at the initiation step depends on the nature of the radicals formed: formation of $\text{R}^{\cdot}$, $\text{RO}^{\cdot}$, and $^{\cdot}\text{OH}$ radicals results in oxygen consumption, whereas formation of $\text{ROO}^{\cdot}$ rules out such a possibility. The rate of oxygen consumption in the initiation events can in general form be expressed as nax where $n = s/l$, $0 \leq n \leq 2$.

Expressing the concentration of peroxy radicals according to equation (3) and introducing a designation $b = k_p[\text{RH}](k_i[\text{Ct}]_0/k_t)^{1/2}$, we can put down the equation for the rate of hydroperoxide accumulation (W_H) in such a process:

$$W_H = b(mx)^{1/2} - ax \tag{4}$$

The oxidation rate W, expressed through the oxygen consumption rate, can under steady-state conditions be presented as

$$W = \frac{d[O_2]}{dt} = k_p[\text{ROO}^{\bullet}][\text{RH}]$$
$$+nk_i[\text{Ct}]_0[\text{ROOH}] - 1/2\,k_t[\text{ROO}^{\bullet}]^2;$$

and , using the designations adopted above, it can be put down as

$$W = b(mx)^{1/2} + \left(n - \frac{m}{2}\right)ax \tag{5}$$

Let us analyze equations (4) and (5). A condition for attaining the maximal possible rate of hydroperoxide accumulation (W_{Hm}) is the equality $dW_H/dx = 0$, i.e.,

$$\frac{dW_H}{dx} = \frac{1}{2} bm^{1/2} \cdot x^{-1/2} - a = 0.$$

Hence it follows that W_{Hm} corresponds to a hydroperoxide concentration x_{Hm} equal to

$$x_{Hm} = \frac{mb^2}{4a^2} \tag{6}$$

A condition for attaining the steady-state oxidation rate W_s and the maximal hydroperoxide concentration $x_s$is the equality $W_H = 0$; hence

$$x_s = \frac{mb^2}{a^2} \tag{7}$$

The rate conforming to conditions (6) and (7) are obtained by substituting the corresponding values of x in equations (4) and (5):

$$W_{Hm} = \frac{mb^2}{4a} = C_{Hm} \frac{k_p^2 [RH]^2}{k_t} \tag{8}$$

where $C_{Hm} = m/4$, and

$$W_s = (1 + n - m/2) \frac{mb^2}{a} = C_s \frac{k_p^2 [RH]^2}{k_t} \tag{9}$$

where $C_s = (1 + n - m/2)m$.

The experimental kinetic data [1–5] have shown that the oxidation rate passes through a maximum. The maximum can be attained on condition that $dW/dx = 0$, hence

$$x_m = \frac{mb^2}{4(m/2 - n)^2 \cdot a^2}, \tag{10}$$

$$W_m = \frac{mb^2}{4(m/2 - n)a^2} = C_m \frac{k_p^2 [RH]^2}{k_t} \tag{11}$$

where

$$C_m = \frac{m}{4(m/2 - n)}$$

Thus, for oxidation initiated through catalytic decomposition of hydroperoxide, coefficient C_s in equation (9) can be equal not only to 1 or 4 as predicted by the classical mechanism [6]. From equations (9) and (11) it also follows that depending on the initiation mechanism the values of C_s and C_m do not always coincide. Table 1 presents the possible values of parameters C for the initiation mechanisms considered above (i–iii) as well as for other putative mechanisms at boundary values of m and n. According to equation (5), at $n \geq m/2$ the rate must monotonically increase and has no extremum in the region of positive x values. At $n = 0$, i.e. when oxygen consumption is only associated with hydroperoxide formation during chain propagation but does not take place at the initiation step, the oxidation rate may pass through a maximum (Table 1, cases iii and v); however, the values $C_m = 1/2$ thus obtained differ markedly from those ($C_m \approx 1/9$) observed by us for cumene [2] and acenaphthene [3] oxidation.

Furthermore, the condition of a steady-state when $W_H = 0$ must drastically diminish selectivety for hydroperoxide. Yet, according to our data [2], when chlorobenzene is used as a solvent, the hydroperoxide yield per consumed oxygen in most experiments remains high (over 90%) throughout the reaction, which means that in our hands there is no steady state for hydroperoxide in the reaction.

The above analysis demonstrates that the values of W_{Hm}, W_s, and W_m are in all cases proportional to $[RH]^2$. This allows an important generalization. Let us show that if chain termination is quadratic and if the functional dependences of initiation and hydroperoxide consumption rates on the concentration of hydroperoxide and catalyst are similar and equal to $f(x)$, then the rate will remain proportional to the square of the concentration of the oxidized hydrocarbon.

We shall rewrite equations (4) and (5) as

$$W_H = bm^{1/2}(f(x))^{1/2} - a \cdot f(x) \quad (12)$$

$$W = bm^{1/2}(f(x))^{1/2} + (n - m/n) \cdot a \cdot f(x) \quad (13)$$

where $f(x)$ is the functional dependence of the hydroperoxide decomposition rate on its concentration and the catalyst concentration.

Then we shall differentiate equations (12) and (13):

$$\frac{dW_H}{dx} = \frac{1}{2} bm^{1/2} \frac{f'(x)}{(f(x))^{1/2}} - a \cdot f'(x) \tag{14}$$

$$\frac{dW}{dx} = \frac{1}{2} bm^{1/2} \frac{f'(x)}{(f(x))^{1/2}} + (n - m/2) \cdot a \cdot f'(x) \tag{15}$$

Table 1. Kinetic parameters C in the reaction rate equation (1) for different initiation mechanisms.

Type	m	n	C_{Hm}	C_s	C_m
i	2	2	1/2	4	–1/2
ii	1	1/2	1/4	1	∞
iii	2/3	0	1/6	4/9	1/2
iv	2	1	1/2	2	∞
v	2	0	1/2	0	1/2
vi	1	0	1/4	1/2	1/2

Hence we obtain the expressions:

$$f(x)_{Hm} = \frac{mb^2}{4a^2} \tag{16}$$

$$f(x)_s = \frac{mb^2}{a^2} \tag{17}$$

$$f(x)_m = \frac{mb^2}{4(m/2 - n)^2 a^2} \tag{18}$$

If $f(x) = x$, expressions (16) to (18) convert respectively into (6), (7), and (10). Thus, formally the values of the characteristic rates W_{Hm}, W_s, and W_m are independent of the type of functions $f(x)$ and of the nature of the catalyst, and are constants for a given substance at the chosen temperature, provided that the initiation and the termination mechanisms are maintained. However, since the type of function $f(x)$ may change substantially with the properties of the catalyst, at $f(x) < x$ the characteristic rates would be attained at higher hydroperoxide concentrations as compared with the case when $f(x) = x$.

5. Catalyst Consumption

The above analysis does not explain the observed decline in the rate after reaching a maximum at small consumption of the oxidized substance [2], while the theory predicts a steady state [6], and does not explain the dependence of W_m on the initial concentration of the catalyst [4] and the hydroperoxide [5]. One of the most plausible causes of this disparity is that catalyst consumption has been neglected in the analysis of the oxidation scheme, and therefore the catalyst concentration does not enter into the expressions for the characteristic rates. Under real condition the catalyst may be completely deactivated before the system accumulates the hydroperoxide in amounts ensuring the characteristic values of $f(x)$ and correspondingly W_{Hm}, W_s, or W_m.

For quantitative assessment of the correspondence of this suggestion with the experimental data, computer simulation has been carried out for the kinetics of oxygen consumption and hydroperoxide accumulation according to the simplest of all possible schemes taking into account the consumption of the catalyst in the reaction with the hydroperoxide:

$$\frac{dC_1}{dt} = k_1\sqrt{C_1C_2} - k_2C_1C_2$$

$$\frac{dC_2}{dt} = -k_3C_1C_2 \qquad (19)$$

$$\frac{dC_3}{dt} = k_1\sqrt{C_1C_2} + k_4C_1C_2$$

where C_1, C_2, and C_3 are the concentrations (moles/liter) of cumene hydroperoxide, catalyst, and the amount of consumed oxygen, respectively; and kj are the constants for given experimental conditions (temperature, cumene concentration) that include the rate constants for the corresponding reactions.

The results thus obtained are presented in Tables 2 and 3. The calculations show that not all constants k_j (with the same j) remain the same for any experiment within one series. At the same time, analysis of the dependence of k_1 on the catalyst concentration at constant RH concentration reveals a tight functional link between these two values. This relation is best described by functions with inverse proportionality; in particular the function $k_1 = 1/(A \cdot C_2 + B)$ displays a correlation coefficient of 0.99.

Table 2. Kinetic parameters of cumene (RH) oxidation in the presence of n-butyl pyridinium bromide, in benzonitrile as solvent (*benzonitrile–chlorobenzene 1:1 by vol.); t_{ind} is the induction period, t_m is the time of reaching W_m; W_m is the maximal oxidation rate; C_{2o}, C_{1o} are the initial concentrations of n-butyl pyridinium bromide and cumene hydroperoxide, respectively.

Expt. No.	$C_{2o} \cdot 10^3$ moles/liter	RH moles/liter	Experimental t_m min	Experimental $W_m \cdot 10^5$ mol · $l^{-1}sec^{-1}$	Calculated by eqns. (19) t_{ind} min	Calculated by eqns. (19) t_m min	Calculated by eqns. (19) $W_m \cdot 10^5$ mol · $l^{-1}sec^{-1}$
			Series 1: 343 K				
1	0.17	3.58	240	2.47	101	222	2.11
2	0.35	3.58	153	3.26	56.0	144	3.36
3	3.5	3.58	127	3.35	16.0	123	3.38
4	9.0	3.58	108	4.79	16.4	90	4.49
5	12.3	3.58	69	4.44	14.2	84	4.98
6	25.7	3.58	195	2.94	41.1	213	2.90
			Series 2: 343 K, C_{1o}= 0.326M				
7	0.29	3.58	9	4.48	6.74	6	4.56
8	0.058	3.58	6	3.69	5.09	0	4.15
9	0.26	3.58	6	4.79	2.76	0	5.64
10	4.6	3.58	9	6.02	1.77	0	6.17
			Series 3: 358 K				
11	0.07	3.58	30	9.66	22.3	34	9.37
12	2.2	3.58	36	10.8	3.25	42	10.1
13*	4.5	3.58	35	12.3	10.1	44	11.6
14	12.0	3.58	30	11.1	–0.61	42	11.6
15*	4.5	2.87	30	8.04	12.2	46	7.98
16*	4.5	2.15	42	4.59	11.0	54	4.49
17*	4.5	1.43	90	1.94	25.2	93	1.91

Table 3. Kinetic parameters calculated from equation (19). W_{Hm} is the maximal hydroperoxide accumulation rate, S_{O_2} and S_H are the sum of square derivatives for oxygen consumption and hydroperoxide accumulation. Experiments are numbered as in Table 2.

Expt. No.	$W_m \cdot 10^5$ mol · $l^{-1}sec^{-1}$	k_1 sec^{-1}	k_2	k_3 liter · mol^{-1} · sec	k_4	$S_{O_2} \cdot 10^3$	$S_H \cdot 10^3$
1	1.40	0.10	16	0.14	35	3.1	2.4
2	2.31	0.10	8.3	0.17	28	1.6	1.6
3	2.27	0.037	2.6	0.12	0.43	2.4	1.7
4	2.99	0.026	0.53	0.18	1.2	1.6	1.7
5	3.32	0.024	0.37	0.21	1.1	2.9	1.6
6	1.87	0.0096	0.23	0.050	0.050	1.8	1.9
7	3.07	0.22	56	0.16	38	2.3	2.1
8	2.90	0.12	10	0.20	30	1.8	1.7
9	3.86	0.10	7.7	0.10	5.0	4.7	2.9
10	4.29	0.02	0.11	0.085	0.64	1.8	2.8
11	6.60	1.2	674	4.0	981	2.7	2.4
12	6.37	0.13	14	0.30	1.9	2.0	6.1
13	7.68	0.10	5.5	0.38	2.9	1.3	2.4
14	7.16	0.049	1.0	0.26	2.0	5.0	5.0
15	5.50	0.10	4.4	0.45	2.4	1.0	1.2
16	3.12	0.094	4.4	0.45	0.48	0.61	0.83
17	1.38	0.070	2.0	0.46	0.92	0.78	0.76

Experiments 13, 15–17 (Table 2) with varying cumene concentration, when processed by a logarithmic anamorphosis of W_m *vs.* [RH], exhibit good conformity of the maximal rate to equation (1), since the reaction order for the substrate proves close to 2 both for experimental (Figure 2) and calculated W_m (Tables 2 and 3). It is noteworthy that, according to the calculations of catalyst consumption, for all experiments (except expts. 7–10 with initial hydroperoxide additions) the maximal rate is reached at the moment when the catalyst concentration in the mixture falls by a factor of 2–2.5. These results show that equation (1) may hold true not only in the classical cases of oxidation initiated through hydroperoxide decomposition under the action of a catalyst without any expenditure of the latter [6,7]. The simple model considered in the present work exemplifies that such a dependence may be observed also in the cases when the catalyst is spent in the course of oxidation. Thus, the linear dependence between W_m and the square of hydrocarbon concentration, which is widely used as the criterion of applicability of the classical scheme (not accounting for catalyst consumption), is not a sufficient criterion.

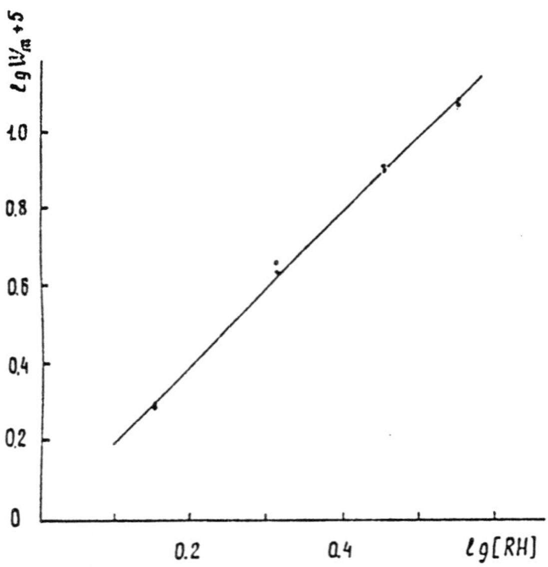

Figure 2. Logarithmic plot of the dependence of the maximal oxidation rate on cumene concentration at 358K. $C_{2o} \cdot 10^3$ (mole/l); the solvent is benzonitirle with chlorobenzene (the volume portion of benzonitrile in solution is 0.5).

Simulation of the process without accounting for catalyst consumption, i.e. by the simple scheme [6], demonstrates for the same kinetic curves that there are no positive k values with which the calculated time of attaining the maximal oxygen consumption rate would coincide with the one observed experimentally, when hydrocarbon consumption only amounts to a few percent and cannot be the cause of the subsequent decline in the oxidation rate. This is another proof of the inadequacy of the simple scheme, and of the need to take into account additional reactions such as catalyst consumption. The impossibility of describing the shape of the kinetic curve of oxygen consumption without accounting for catalyst consumption is an important criterion of the invalidity of the classical scheme in this case.

No definite conclusions about the mechanism of initiation involving a catalyst can be made from the data of the present study which has been basically aimed at a formal kinetic analysis of the oxidation process as a whole. Still, the fact that there is a nice correlation between k_1 and C_2 and that they are in inverse proportion may indicate that the catalyst is involved in initiation preferentially in the ionic form; in this case the concentration of active particles—ions—would be proportional to the square root of C_2, which would result in an inversely proportional dependence of the effective k_1 on C_2.

The available experimental data on oxygen consumption and hydroperoxide accumulation, and the calculated data on catalyst consumption make it possible to assess the catalyst efficiency in the systems studied (the amount of oxygen consumed or hydroperoxide formed per unit concentration of catalyst spent). Such evaluation demonstrates that with a decrease of the initial catalyst concentration its efficiency rises and may reach values of the order 10^3, i.e. may be quite high. As to the question in which reaction the catalyst is consumed and what attentuates its efficiency, it can only be answered by using more detailed schemes for simulation and designing special kinetic experiments that would explore the process to deeper extents of conversion. The fact that constant k_3 does not greatly change in all curves simulated (Table 3) points to that the catalyst is consumed in a reaction in which it takes part in the molecular form.

6. Conclusions

(1) The kinetics of cumene oxidation to hydroperoxide has been studied in the presence of one of the most active onium salts, n-butyl pyridinium bromide.

(2) Formal kinetic analysis and computer simulation have been carried out for the process of hydrocarbon oxidation to hydroperoxide in the presence of a catalyst, in the case when initiation takes place through the reaction of the end product—hydroperoxide—with the catalyst. It is shown that the consumption of the catalyst in the course of oxidation must be taken into account at small extents of conversion of the initial hydrocarbon.

(3) It is shown that the proportionality of the maximal oxidation rate to the square of hydrocarbon concentration is not a sufficient criterion of the validity of the classical oxidation scheme, because this condition may also be fulfilled in the case of catalyst consumption.

(4) Analysis of the kinetic parameters calculated using the proposed mathematical model allows a suggestion that the catalyst takes part in the consumption reaction in the molecular form. As to the initiation step, the inverse dependence between the parameter characterizing this reaction and the catalyst concentration allows a suggestion that in the initiation process the catalyst participated in the ionic form. More exact conclusions concerning this question require more detailed studies of this reaction.

7. REFERENCES

1. **I.A. Opeĭda, N.M. Zalevskaya, N.I. Mironenko,** *Ukrainskiĭ Khimicheskiĭ Zhurnal,* Vol. 52, No. 8, 1986, pp 884–889, Russian.
2. **I.A. Opeida, N.M. Zalevskaya,** *Neftekhimiya,* Vol. 27, No. 5, 1987, pp. 678–685, Russian.
3. **I.A. Opeĭda, N.M. Zalevskaya, S.A. Gerasimova, A.G. Matvienko,** *Neftekhimiya,* Vol. 28, No. 4, 1988, pp. 544–551, Russian.
4. **I.A. Opeĭda, N.M. Zalevskaya,** *Neftekimiya,* Vol. 29, No. 2, 1989, pp. 244–250, Russian.
5. **I.A. Opeĭda, N.M. Zalevskaya,** *Ukrainskiĭ Khimicheskiĭ Zhurnal,* Vol. 56, No. 1, 1990, pp. 79–82, Russian.
6. **A.V. Tobolsky, D.J. Metz, R.B. Mesrobian,** *J. Am. Chem. Soc.,* Vol. 72, No. 5, 1950, pp. 1942–1952.
7. **N.M. Emanuel, E.T. Densiov, Z.K. Maizus,** *Tsepnye Reaktsii Okisleniya Uglevodorodov v Zhidkoi Faze* (Chain Reactions of Hydrocarbon Oxidation in Liquid Phase) Moscow: Nauka, 1965, 329 pp., Russian.
8. *Technique of Organic Chemistry, VII Organic Solvents* (A. Weissberger), New York: Interscience Publ., Inc., 1955, 519 pp.

SUBJECT INDEX